D0425966

BRASSES AND BRASS RUBBING

Clare Gittings

London
Blandford Press

First published 1970 by
Blandford Press Ltd.,
167 High Holborn,
London WC1V 6PH
Second impression, 1971

ISBN 0 7137 0520 5

Printed in Great Britain
by Fletcher & Son Ltd, Norwich

CONTENTS

DEDICATION

To Dr. A. C. Bouquet and
Jerome Bertram

ACKNOWLEDGEMENTS

I should like to thank Dr A. C. Bouquet, author of the magnificent *European Brasses* (Batsford 1967) for his friendship and help with this book, and also Jerome Bertram for reading the typescript and showing me so many of his own rubbings and for supplying one of the illustrations. I should also like to thank my parents, Robert Gittings and Jo Manton: my father for typing the manuscript and transporting me; my mother for making several helpful criticisms. Thanks should also be given to Charles Saumarez Smith and Malcolm Ramsay for reading the proofs. Thanks are also due to R. C. H. Briggs, Honorary Secretary of the William Morris Society, for information, and to the Society of Antiquaries for permission to photograph and reproduce the rubbing at Kelmscott Manor House; and to the Rectors of Edlesborough (Bucks.) and Trotton (Sussex) for rubbing permissions not normally granted. I must thank Dr. G. H. S. Bushnell for much help, and for the loan of a brass from the University Museum of Archæology and Ethnology, Cambridge; the *Evening Argus*, Brighton, for permission to use the photograph on page 17. Finally I should especially like to thank the clergy of all the parishes where I have rubbed, without whose kind co-operation this book would not have been possible, and in particular the Rev. C. E. Matthews for his permission to reproduce the rubbing shown in fig. 32.

1. *Cover:* Sir Roger de Trumpington
1289, Trumpington, Cambs.

MONUMENTAL BRASSES

About 8,000 monumental brasses exist in churches in the British Isles, though these represent only a small proportion of the total number laid down between about 1250 and 1650. Most people have seen monumental brasses in a church: flat metal figures engraved with lines to show the features.

A brass memorial had three main advantages over stone or alabaster figures. A brass was generally cheaper, and if there was not enough money for a whole figure brass, a chalice, heart or symbol could be laid down. A brass memorial did not take up as much room as a stone tomb; brasses are often found paving the aisle of a church. The third advantage was that brasses were not as easily defaced as stone figures. There are very few stone monuments still perfect; most have lost their noses or hands, and have been scrawled with initials, Brasses, being so much harder than stone, have not often been damaged in this way.

A great many churches in England contain brasses, and if there are none in your own church, there will probably be some in a church near by. You can make a copy of these brasses most effectively by brass-rubbing. With other pieces of church ornament, such as stained glass or carved stone, it is impossible to make an exact copy without losing some of the charm and clarity of the original. Often, however, a brass-rubbing is clearer than the actual brass and, besides being a valuable historical record, it is a small work of art.

Part One: HOW TO MAKE AND COLLECT BRASS RUBBINGS

BRASS RUBBING: INTRODUCTION AND EARLY HISTORY

Brass-rubbing is a comparatively easy and cheap hobby; beginners can become expert at it in a relatively short time. It is very like rubbing a penny with a pencil; you place a paper over the brass, rub with a wax stick, and the design appears on the paper, the engraved lines remaining white.

Brass-rubbing became popular as the hobby of the Victorians, particularly clergymen, who would hang up their rubbings on the very high walls of their Gothic vicarages. An earlier method of reproducing brasses consisted of pouring printers' ink into the engraved lines, and then placing a sheet of damp tissue paper over the brass. This gave a very poor impression; besides, the design was reversed, so that inscriptions became unreadable. In any case, very few incumbents would allow this method of copying brasses.

A black shoemakers' wax, heelball, then came into use, and it is still the best wax for making rubbings. William Morris, the designer of fabrics, wallpaper and books, records in a letter headed 'Tuesday in Holy Week' and dated 1856, when he was twenty-two:

> The other day I went 'a-brassing' near the Thames on the Essex side; I got two remarkable brasses and three or four others that were not remarkable: one was a Flemish brass of a knight, date 1370, very small: another a brass (very small, with the legend gone) of a priest in his shroud; I think there are only two other shrouded brasses in England.

The letter was written to Cormell Price from Walthamstow, in Essex, and these 'remarkable' brasses are from Aveley and Stifford, both near the Thames and only a few miles away from each other. The Aveley brass is a small foreign plate to Ralph de Knevyngton, and the Stifford brass is of a priest in a shroud holding an inscribed heart. Morris is, of course, wrong in thinking that there are only three shroud brasses in England. He also mentions that he is planning to go to Stoke d'Abernon, where there is the oldest brass in England. When he moved later to his famous house, Kelmscott Manor in Berkshire, Morris rubbed the brass at Great Coxwell in

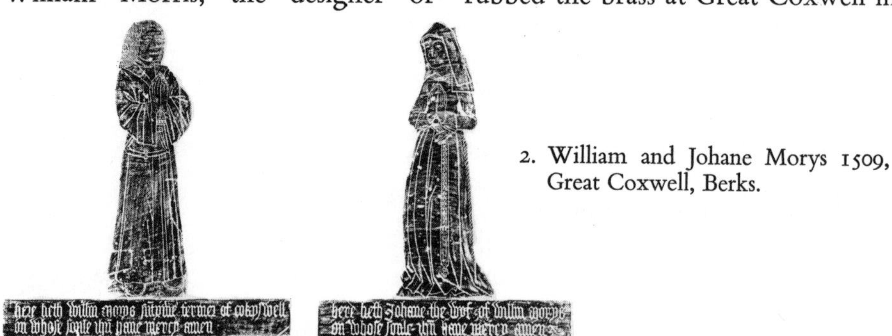

2. William and Johane Morys 1509, Great Coxwell, Berks.

the same county. This had a particular interest for him, as it was to William Morys, 'sumtyme fermer of Cokyswell', who died in 1509. A rubbing of this brass, illustrated here, still exists and may be seen at Kelmscott. Nearer our own time, Lawrence of Arabia also rubbed brasses near Oxford. The methods they employed for rubbing are used today, helped by such inventions as Sellotape or masking tape to keep the paper still, rather than church hassocks or books.

In the British Museum is the collection by Craven Ord, the first collector of rubbings from 1780 to 1830, made by the printers' ink method. There are also a number of 'dabbings', made by placing tissue paper over the brass, and then dabbing it with a soft leather cloth, covered with a paste of Chinese ink or powdered black-lead mixed with linseed oil. This method is still in use when a brass is engraved with very shallow lines, as in the late copper plates.

In both the British Museum and the Victoria and Albert Museum, there are large collections of rubbings. These rubbings are not on general display, but may be seen if due notice is given.

Several books have been written about brasses and brass-rubbing, mostly in the 19th and the early part of the 20th centuries. Most of them, such as H. Haines's *A Manual of Monumental Brasses* (1861) and Herbert W. Macklin's *The Brasses of England* (Methuen 1907), are now out of print; but they can sometimes be found in second-hand book shops. With the present awakening interest in brasses and brass-rubbing, several new books have been published, of which Malcolm Norris's *Brass Rubbing* (Studio Vista, latest edition 1969) is particularly good and has detailed illustrations. Though expensive, Mill Stephenson's *A List of Monumental Brasses in the British Isles* (Monumental Brass Society, reprinted in one volume, 1964) is of the greatest value to any brass-rubber. This large book contains a detailed list of all the existing brasses in Great Britain and Ireland, and is extremely accurate and helpful. Another book, often reprinted, and recently edited and rewritten by John Page Phillips which is a good introduction to brasses, is H. W. Macklin's *Monumental Brasses* (Allen & Unwin, 1970). For information about brasses in an individual county, there are usually articles on the subject in the county's Magazine or Archaeological Proceedings, and these may be found in a good local reference library. Brasses on the Continent suffered greatly during the First and Second World Wars, and as yet no one has fully revised W. F. Creeny's book on Continental Brasses, published in 1884.

In 1887, a society was formed to care for and promote interest in monumental brasses. It has since become known as the Monumental Brass Society; for details of membership see page 101.

MATERIALS FOR BRASS-RUBBING

You will not need many materials for brass-rubbing, and they are quite easy and cheap to obtain. The following is a list of all that is needed to make a rubbing: heelball, paper, masking or Sellotape, scissors, two dusters, soft nail-brush, notebook and pencil.

Heelball is a hard black wax, sold usually in sticks. There are various types of heelball for sale, most of which may be obtained from a good art store. The Monumental Brass Society recommends a brand called Astral. I use Summitwax, which is a little softer, so that one does not need so much pressure to make a good rubbing.

When you first begin brass-rubbing, I suggest you use ordinary lining-paper. This is cheap and easy to obtain, and is very useful for beginners. However, it will turn yellow very quickly, and so it is not recommended if you wish to keep the rubbings for any time. When you feel that you have mastered the basic technique of rubbing, you should use a better paper. Architects' detail paper is the best. It is made in a variety of widths: 30 inches, 40 inches, 60 inches and even 80 inches. This has obvious advantages over lining-paper, which is only 22 inches wide. There are many leading art suppliers which stock architects' detail paper, and it can be bought at any good art shop. There are various types and weights of paper, usually indicated by numbers. I use architects' detail paper No. 78, made by Harper & Tunstall, supplied at their trade counter, 39, Victoria Street, London, S.W.1, but it is possible to get good results on slightly different weights of paper.

The masking or Sellotape is for holding down the paper. In some churches you are not allowed to use Sellotape, so it is advisable to have both. Scissors are mainly used for cutting a sufficient length of paper off the main roll, and for snipping the tape. When there is an awkward projection near the brass, such as the supports of an iron railing round it, this difficulty can be overcome by cutting a slit in the paper so that it fits around the obstacle, and can be laid flat.

One of the dusters and the soft nail-brush are for cleaning the brass before you begin. This is very important indeed. The other duster is for polishing the rubbing when you have finished it. The notebooks and pencil are for noting down any details about the brass and the church.

Several variations on the normal black-on-white rubbing can be made. You can use gold wax on black paper, black wax on gold paper, or any other combination of colours. For special details, mainly heraldic devices, it is a good idea to have a set of coloured waxes, which are easy to buy. A German firm manufactures Colorex sticks in silver and several shades of gold, costing only a small amount, which can be most effectively used for heraldic tabards, shields and linings of costume. The use of coloured waxes and papers is a matter of taste, but I do not recommend gold wax used on black paper, as it gives a very blurred result. On the other hand, an ordinary black heelball rubbing done on gold paper gives a most pleasing effect; it retains the clarity of a familiar black-on-white rubbing, while adding a richness of texture. Several wallpaper firms make the plain gold paper, of 22 inch width, which is sold at a reasonable price. A more expensive gold paper of 30-inch width can also sometimes be obtained. Such materials, however, are really for the specialist; the beginner will be best equipped with the basic list I have suggested.

ASKING PERMISSION

When you have decided in which church you want to take a rubbing, and have bought the necessary materials, you must get permission to rub from the vicar or rector – or in the case of a Cathedral, the Dean. The brasses are their responsibility, and they can grant or refuse permission as they wish. It is stated in the *Opinions of the Legal Board of the National Assembly of the Church of England* (4th edition, p. 117) that: 'The permission of the incumbent to the taking of rubbings in a church is required by right of his freehold, and he can make it conditional on the payment of a fee.'

The incumbent's name, address and telephone number can be found in Crockford's *Clerical Directory*, which can be consulted at most libraries. By looking up the parish in the second half of this book, you will find the incumbent's name and initials; then turn to the first half, and look him up in the alphabetical list of clergy. Among other details, it will give you his correct address and telephone number. Most clergy are very busy people so write to him well in advance for permission to rub, giving, if possible, alternative dates on which you could come, and enclosing a stamped, addressed envelope.

If there is not time to write, you can telephone the incumbent. Ask to speak to the vicar or rector, as the case may be (Crockford's will indicate which he is by 'V' or 'R'). Be brief and very clear about the day on which you want to rub, as he will probably have to check that no one else is rubbing at that time.

If you are not able to consult Crockford's or get in touch in advance, you can arrive without notice at the church, but often you may not be able to rub. You may find other people rubbing, or that permission is not given at all. It is always wise to look in the church first to see if there are any rubbing regulations. In any case, you should go and find the rector or vicar. His name and address will probably be on the church notice board, and you should ask him personally for permission. If you cannot find him or a churchwarden, who may also sometimes give permission, then you should not rub the brass.

Many clergy will not allow rubbing at week-ends, as the church is usually cleaned on Saturdays ready for the Sunday services. Most clergy will charge a fee for rubbing, as they are entitled to do, though some only ask you to put what you like in the church box. The fee is generally quite moderate and when a larger fee is charged, there is often a reduced rate for students and young people. Sometimes you will be asked to sign a form saying that you will not use the rubbing for commercial purposes or profit. You will generally find that parish clergy are very helpful and interested in your rubbing, provided you help them by following these rules.

HOW TO RUB

Having got permission to rub, you should then go to the church in good time to look for the brass. This may not sound difficult, but in fact brasses are often laid or relaid in places where you might not normally think of looking: on walls, in the vestry, under large carpets or sometimes floor-boards, or even fixed to the back of the organ. It can take quite a long time to find them, though directions in church guide-books will often help.

The first thing to do is to inspect both the brass and the stone slab in which it is set. This is usually firm, but if the stone is crumbling, it may be necessary to use weights – such as the old device of church hassocks – instead of masking or Sellotape, for fixing down the paper. Then clean the whole brass with a duster and soft nail-brush. This is most important, as a single speck of grit can cause a rubbing to tear.

Study carefully all the details of the brass. Many brasses have intricate out-lines, and it can be all too easy to leave a bit out by mistake. It is a good idea to draw a plan of the brass, showing clearly such things as the hilt of a sword, a dagger, and any stray pieces of the canopy. Sometimes there is a picture or a postcard of the brass in the church, and this is a very helpful guide. At Wiston (Sussex) the brass to Sir John Braose contains over thirty small scrolls saying 'Jesu' and 'Mercy'; you certainly need a very accurate plan to avoid missing any of them.

Next, stick down the paper over the whole brass, fastening it to the stone surround with tape or weights. The paper must be absolutely flat before you begin to rub, and must be really firmly stuck down. Loose paper will give a blurred rubbing, and many lines and details will be lost.

A point often forgotten is that heel-ball can become extremely cold and hard, particularly in winter; it is a good idea to soften it a little before you begin rubbing. This will stop it flaking off. Polish the stick with a duster for a few minutes until it starts to shine. Then begin to rub with the heelball over the centre of the brass. The rubbing should be an even black with clear white lines. You should rub carefully and methodically, putting as much pressure as you can on to the heelball; I suggest that you begin by choosing a brass which is laid on the floor, as the weight of your shoulders will help to give more pressure. When you reach the edges, be very careful not to go over them and rub the stone, as this will spoil the outline. To begin with, use the thumb of your other hand as a barrier to stop you going over the edge, and later you will probably develop your own way of rub-bing edges.

It does not really matter in what direction you rub, but it is advisable to rub across, and not along the lines of engraving. Sharp edges of metal, especially on the earlier brasses, may

3. The author demonstrating how to rub, using a brass to a post-Reformation priest *c.* 1610

A. Cleaning the brass with a soft duster before rubbing

B. Sticking down paper over the brass with masking tape, making sure the brass is centrally placed under the paper

C. Starting to rub a central part of the design

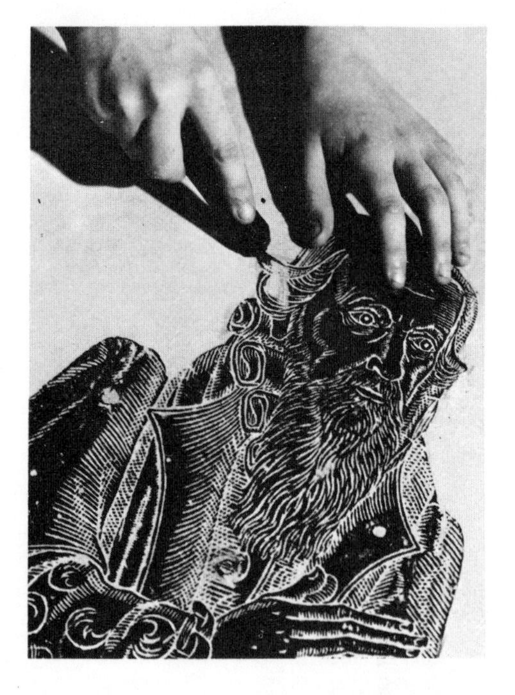

D. Rubbing detail at edge of brass, holding paper firmly down with the other hand

E. Filling in general design of the brass. Note pressure of index finger, with shoulder of rubbing arm well over the brass

F. The completed rubbing, with flakes of loose heelball dusted off, ready for removing masking tape and rolling up the paper

make you tear the paper in spite of all precautions. Rivets, too, can tear the paper, and these should be carefully noted before rubbing. These tears can easily be mended later with paper and paste on the back of the rubbing. Minor marks of heelball over the edge can also be removed later with an eraser.

When you think that you have rubbed every part of the brass, first make sure that you have not left out anything, especially in a canopy over the figure. Look at the rubbing from all sides to see that the heelball is an even black all over; take special care that any inscription is fully legible. When you are satisfied that the rubbing is completed, polish it with a soft cloth, then take it up, being careful to remove all the Sellotape or any fastenings first. Put back all the church furnishings exactly where you found them, and then roll up the rubbing, wiping any loose flakes of heelball off, and take care to remove these from the church.

Always leave yourself plenty of time to do a rubbing; if you hurry you will get a bad result. After a while you will discover at what speed you rub, and you can plan your time accordingly. Never try to rub too many brasses in a short space of time; one good rubbing is easily worth six hurried ones.

OTHER THINGS TO RUB

In recent years the rubbing of brasses in churches has become increasingly popular; it is not generally realized that the technique of brass-rubbing can be used in other fields as well.

4. Two horse brasses

Many people already rub manhole covers and iron firebacks, and there are numerous other suitable objects such as horse-brasses, graffiti (scratchings on walls) or even different types of embossed wallpaper. Decorative rubbings of these kinds can be made in any country.

An interesting type of rubbing, though rare, is that of inscriptions round church bells. A few bells have been removed from their belfries, and placed on the floor of the church. These often show the names and dates of the bell-founders, or the donors, or of the vicar and churchwardens of that time. It is, of course, also necessary to obtain permission from the incumbent to rub these. The bell at Caversfield (Oxon.), the oldest inscribed bell in England, is dedicated to St Laurence, and names the donors.

One can also rub inscriptions on slate tombstones. While I was in America, I made quite a large collection of unusual tombstone-rubbings. For example, Mary Goose from the Granary Burial Ground, Boston, was the original 'Mother Goose', who made up nursery rhymes for her children. In the same place are inscriptions

5. Inscribed bell *c.* 1200, Caversfield, Oxon.

6. Mary Goose 1690, Granary Burial Ground, Boston, Mass., U.S.A.

to many of the famous Boston patriots, such as Paul Revere. At Lexington, Massachusetts, there are sets of tombstones with strange hollow faces engraved on them, very like primitive African art. Incidentally, when you are rubbing out of doors, it is as well to choose a fine day. One of my tombstone-rubbings blew away over a church and houses, and landed in someone's garden.

STORAGE, LABELLING, DISPLAY, ETC.

A collection of rubbings is very bulky, and presents various storage problems. The rubbings must not be folded, but keeping large rubbings flat is not practical. However, the advantage of heelball is that it will not smudge or blur, so that if the rubbings can be arranged to take up a small space without creasing, they will remain perfect. The most suitable method of storage is to roll each rubbing lightly into a cylinder shape, and put it inside a cardboard roll. Rolling rubbings so that they do not fold or crease takes practice, and at first it is better to roll the rubbing round something, such as a wooden rod, which can be slipped out and removed, before putting the rubbing inside the cardboard roll. These cardboard rolls can be obtained from any draper, who will probably be very glad to get rid of them. Such rolls, which originally had material wrapped round them, follow the usual widths of material, and are generally 36 or 45 inches. Much longer rolls, up to 8 feet, may be obtained from stores which sell linoleum. Chose a roll a little longer than the width of your rubbing, insert the rubbing carefully, and put a stopper – say, a rolled-up ball of paper – in each end.

Each roll should then immediately be labelled with the place-name, church, name of subject, and date. The labels can either be gummed on to the rolls or attached at one end. When you have made a large collection of rubbings, you will find that this labelling will save hours of searching and unrolling.

Brass-rubbings make very beautiful wall decorations. Smaller rubbings can be framed and hung like a picture. Larger rubbings look most effective if hung on dowelling rods with a bright coloured cord, red or gold. Rubbings on detail paper should be hung by sticking a length of $\frac{1}{4}$-inch-thick dowelling on to each end. Cut two pieces of the dowelling to a little over the width of the paper, and roll the ends of the paper round them twice, fixing with strong glue. The cord can then be passed through eye-screws, screwed into each end of the dowelling. Care should be taken in sticking rubbings directly on to walls. If Sellotape is used, it may remove the plaster, and can also stain the rubbing. Wherever possible the dowelling method should be used, as it will preserve both walls and rubbing.

For public exhibitions, lasting a short time, it is, however, best to use Sellotape to stick the rubbings on large stands of whitewashed softboard. They can also be pinned with drawing-pins, care being taken not to tear the edges.

An exhibition has to be very carefully planned, so as to give a balanced effect. The large and small rubbings should be evenly spaced; for a general exhibition, it is better to have a mixture of types of rubbing rather than, say, all the civilian brasses in one section, and all the knights in another. Each rubbing should be accompanied by brief notes on the people commemorated by the brass; that is, their names, date of death, and any interesting facts about their life, as well as the church, place and county in which the brass is situated. Where possible, general notes on armour, costume, the history of brasses, etc. are much appreciated, and will save you from having to answer many repeated questions.

7. Lady Elizabeth Say 1473, Broxbourne, Herts.

8. Author with rubbing of Sir Robert de Grey 1387, Rotherfield Greys, Oxon.

LITERARY AND HISTORICAL INTEREST

It may be asked what is the use of a collection of brass-rubbings, apart from the pleasure it gives to the owner. Monumental brasses can have great literary and historical interest, and be extremely useful in the fields of education and art.

To pursue this interest, a great deal can often be found out about the people commemorated in a brass from a *County History*. A famous person may be mentioned in the *Dictionary of National Biography*, a copy of which is in most reference libraries.

To show how much can be found out, I have selected a set of brasses commemorating people who appear as characters in Shakespeare's History Plays. The information used is mainly from church guides, the *Dictionary of National Biography*, and, of course, the plays of Shakespeare themselves. There are altogether about ten brasses in England connected in this way with Shakespeare, and I have used these as material for lecturing. Here I have taken two characters represented in monumental brasses, and written out briefly as much as is known about their lives and deaths in the type of abbreviated note-form that I find most useful for exhibitions or for lectures:

Westminster Abbey: Eleanor de Bohun, died 1399. Daughter of Humphrey de Bohun, Earl of Hereford, Essex and Northampton, and Constable of England. In 1374, married Thomas of Woodstock, youngest son of Edward III, Duke

9. William and Margaret Catesby 1494, Ashby St Ledgers, Northants.

of Gloucester and later Earl of Essex and Buckingham. On 10 July 1397, Thomas imprisoned at Calais for alleged treason against Richard II. Died mysteriously, about September of that year. Eleanor left widow with one son and three daughters. Made will at Pleshy Castle, 9 August 1399. Became nun at Barking Abbey. Died of grief over son's death, on 3 October. Buried in St Edmund's Chapel, Westminster Abbey, beside husband, whose brass is now gone. In reign of Henry IV, a man named John Halle confessed that Thomas was smothered with feather pillow by order of Richard II. (*See* page 30.)

Ashby St Ledgers, Northants: William and Margaret Catesby. William executed 1485. William Catesby, fine lawyer, given offices by Duke of Gloucester, later Richard III. Richard very unpopular, and Collyngbourne, later executed for treason, wrote

> The cat, the rat and Lovel our dog
> All rule England under the Hog.

'The cat' is Catesby, and 'the Hog' is Richard III, his standard being a boar.

Catesby made Chamberlain of the Receipt of the Exchequer and, on 30 June 1483, Chancellor of the Exchequer and of the Earldom of March. He was an excellent lawyer, and Lord Rivers, executed by Richard III, made him an executor of his will.

On 22 August 1485, Catesby fought at Bosworth Field. King Richard was killed, Catesby captured, sent to Leicester, and executed three days later. In will, dated 25 August, he leaves all property to his wife, asking her to give back all land wrongfully seized, and divide rest among his sons. His will ended saying he hoped Henry VII would be merciful to them, and begging pardon for his treason. Buried at Ashby St Ledgers, his family home.

Such notes show how, by following up the subject of a brass-rubbing, a great deal of fascinating history can be learnt. Other fields in which brass-rubbing can be used for information are the study of all sorts of costume, particularly medieval, where it has been of great assistance to stage-designers; the development of church vestments; religious art through the centuries; social manners and customs; trades and heraldry.

The second section of this book suggests what you can learn about these topics, and others, from brass-rubbing.

MONUMENTAL BRASSES AND THEIR HISTORY

Brasses originate from incised slabs of stone. These portrayed the same types of design as brasses – figures, canopies, shields, and so on – but not so clearly as brass, because the stone is far less durable and is liable to crumble. The earliest existing brass is dated 1231, but they were made as early as 1200. The oldest English brass at Stoke d'Abernon in Surrey is dated 1277. The oldest brasses are the deepest engraved, and the finest brass-engraving ran in schools of craftsmen, London, Cambridge, York, Ipswich and Bristol being among them. Brasses were also imported from Flanders and Germany; these have a distinctive appearance. Makers' symbols appear on some early brasses; some later brasses are signed.

The material used in the manufacture of brasses is not, in fact, actual brass. It is called latten, and is an alloy containing copper, zinc and a small amount of lead. It was made in large rectangular moulds, and then hammered out by the engraver into the desired shape. The tool used for the engraving is called a burin, and can carve out either very deep or quite thin lines. The lines were sometimes filled with coloured waxes and enamels; where these remain, a rubbing may appear blurred and lacking in detail.

The engraved brass passed into the hands of the mason, who cut a slab of stone (usually Purbeck marble) for it. The outline of the brass was engraved on the stone, then cut away so that the

Part Two: WHAT YOU SEE AND LEARN FROM BRASS RUBBINGS

10. Sir John Dabernoun 1277, Stoke d'Abernon, Surrey

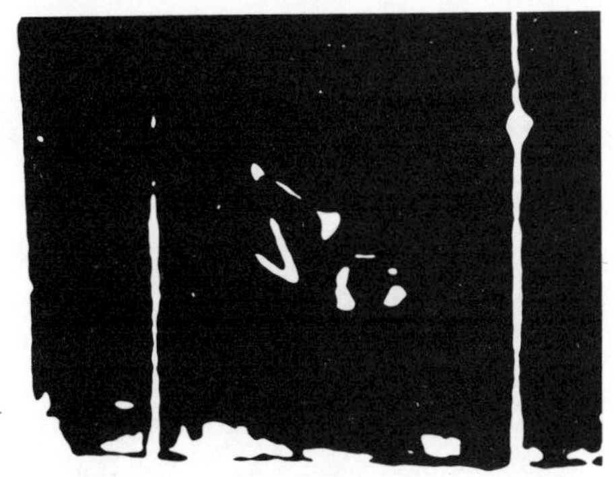

11. Makers mark, detail, 1421, Trotton, Sussex

brass would lie on the same level as the surrounding stone. These indents are called matrices. The matrix was then flooded with pitch, and the brass placed in it. In later times, as well as being held by pitch, the brass was secured by metal rivets. The preservation of these matrices, exposed when the brass has been stolen or lost, is as important as the preservation of the actual brasses, as the matrices give an idea of the shape of the missing brass.

1231 – 1327. The earliest brasses appear between these years. The metal is thick and very deeply engraved. Figures are life-size, and sometimes canopied. Knights are shown cross-legged with a dog or a lion at their feet, and carry shields on their arms. Ladies often have pet dogs playing in the folds of their dress. The inscription is made up of separate letters set round the brass.

1327 – 99. Brasses now become truly magnificent, and often canopied. They range greatly in size, though the engraving is always excellent. The marginal inscription now becomes a continuous strip of metal, and knights no longer carry shields. Cross and bracket brasses of great beauty now appear.

1400 – 60. Brasses are still well engraved, but become smaller. Children start to be shown on separate plates below their parents. Crosses and bracket brasses begin to disappear.

1460 – 85. Brasses now begin to lose their fine engraving. The figures are exaggerated, and shading, in the form of cross-hatching, is used.

1485 – 1558. Brasses are more numerous than ever before, but generally they are badly engraved on thin plates of metal. Single figures are usually given in profile, and mural plates (set on walls) come into fashion. English becomes the language of inscriptions, and a great deal of shading is found. Actual portraits now appear, and increase later. Nicholas

12. Palimpsest inscription 1580, Yealmpton, Devon

Wadham, 1618, is shown with warts on his face (page 36).

1558 – 1650. Engraving becomes very debased. Rectangular plates with little kneeling figures become common. Fewer brasses are engraved and they continue to deteriorate, until the last, in 1773, is merely scratched on a piece of thin bent metal. Many brasses are 'palimpsest' – used more than once. A portion of another figure is found engraved on the back, usually part of a large foreign plate. Brasses are 'reappropriated' – for example, an 18th-century inscription added to a 15th-century brass.

Up to the Reformation, brasses were quite well treated. Henry VIII's commissioners, however, destroyed hundreds, especially monastic brasses. Many were melted down as scrap metal and sold. Edward VI also had thousands destroyed as 'Popish'. Elizabeth, by contrast, attempted to preserve monuments, and passed an act saying they were not to be destroyed, but restored to their churches if possible and mended.

Both the Parliamentarians and the Royalists in the Civil War destroyed thousands of brasses, melting some down to make weapons for the troops. It was then that many cross and bracket brasses were removed, and religious symbols such as the Trinity erased. At Newnham Murren (Oxon.) there are actual bullet-holes in the brass.

During the 18th century, a great many fine brasses were sold or destroyed in rebuilding churches and cathedrals, and again in the 19th

13. Lettice Barnarde 1593, Newnham Murren, Oxon.

century hundreds were lost, destroyed and removed from churches all over England. Many were reset on walls, often very near the roof, at impossible heights for either admiring or rubbing. On the other hand, several brasses were restored during that century very carefully and well.

The present treatment of brasses is quite good, mainly owing to the awakening of interest in brass-rubbing for decoration or historical study. Some instances of damage or bad treatment occur, and have received undue publicity; but for the main part people show much more respect for these historic monuments than their ancestors did. A danger is that their artistic quality has been commercialized, and this has been a handicap to serious brass-rubbers. It is hoped that the genuine study of brasses will overcome this, and contribute to their better preservation in the future.

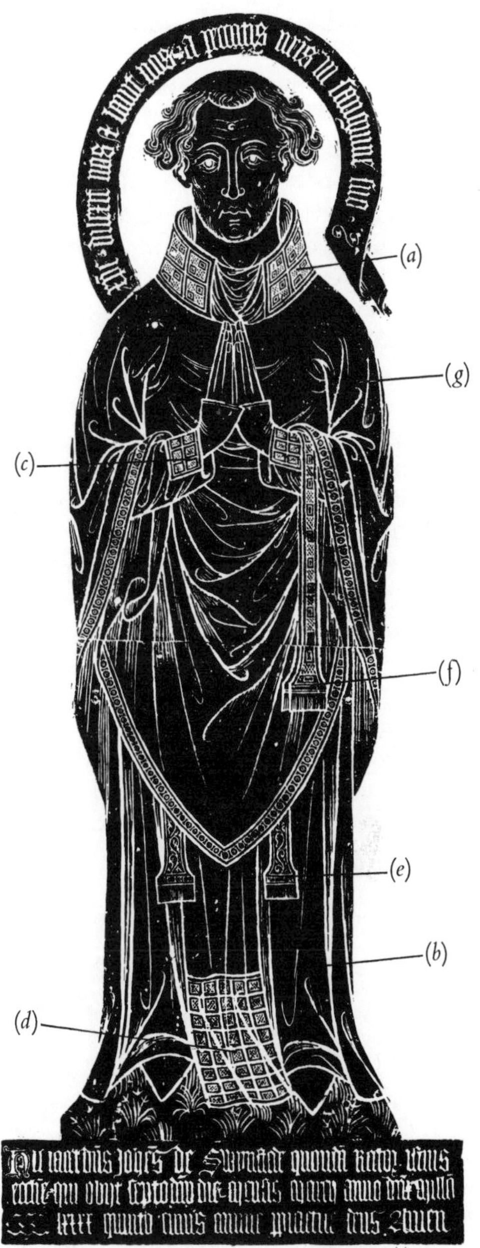

14. John de Swynstede 1395, Edlesborough, Bucks.

MASS VESTMENTS

These are the vestments of a priest celebrating Mass.

- (*a*) Amice
- (*b*) Alb
- (*c*) Sleeve apparels of the alb
- (*d*) Foot apparel of the alb
- (*e*) Ends of the stole
- (*f*) Maniple
- (*g*) Chasuble

The **amice** was a hood, but soon became worn as a neckerchief with an embroidered border called an apparel, usually decorated in gold and silver thread. It appears on brasses like a collar.

The **alb** is a linen vestment reaching to the feet and held by a girdle at the waist. It was adorned with six apparels, only three of which are visible in brass.

The **stole** was a strip of embroidered material passing round the back of the neck and hanging down in front, crossing on the breast.

The **maniple** was, at the time of the Norman Conquest, a napkin for use at Mass, but dwindled down to a silk strip placed over the left wrist.

The **chasuble** is the distinctive mark of the priest. It was an oval vestment with an opening in the centre for the head. It was sometimes decorated with a Y-shaped apparel, especially in larger brasses.

There are numerous brasses showing priests in Mass vestments. The following is a list of most of the best: Higham Ferrers (Northants.), Horsmonden (Kent), North Mymms (Herts.), Shottesbrooke (Berks.) and Wensley (Yorks.). Many of these

15. John Bowthe, Bishop of Exeter, 1478, East Horsley, Surrey

16. Robert Wyvil, Bishop of Salisbury, 1375 Salisbury Cathedral

17. Thomas Cranley, Archbishop of Dublin, 1417, New College, Oxford

brasses are very large with elaborate canopies.

There are many brasses from the 14th century onwards, and these are generally smaller. Some of these figures hold the chalice and wafer in their hands. In Continental brasses these are usually shown lying on the breast, and the hands are crossed, not praying.

BISHOPS AND ARCHBISHOPS

A bishop wears the ordinary Mass vestments of a priest. In addition he wears the dalmatic of a deacon, a fringed tunic reaching below the knee. He also wears the mitre, a pointed jewelled cap with two strips of material, the lappets, hanging down the back. These may be seen in the brass at East Horsley (Surrey). A bishop wears pointed sandals and jewelled white gloves. On his right hand is the episcopal ring. He carries a crosier, a jewelled gold staff with a crook. Good examples occur at the Cathedrals of Carlisle, Ely, Hereford, Manchester and Salisbury, and at Westminster Abbey.

The two brasses of abbots at Westminster Abbey and at St Alban's Abbey show the same vestments. The bishop saints, such as St Nicholas, are also clothed in this way.

Archbishops wear, in addition to these vestments, the Y-shaped pall, and carry a cross-staff instead of a crosier. There are four archbishops in brass, and these are found at York Minster, Westminster Abbey, Chigwell and New College, Oxford. St Thomas of Canterbury quite frequently appears in brasses, wearing the vestments of an archbishop.

PROCESSIONAL VESTMENTS

Many of the finest brasses show priests wearing these vestments:

(*a*) Almuce
(*b*) Cope
(*c*) Morse
(*d*) Orphreys of the cope
(*e*) Surplice
(*f*) Cassock

The **almuce** is a fur cape with a collar and long tails at the front. Sometimes the cope is omitted to show the full almuce. Often traces of enamel can be found on the almuce, and occasionally 'silver' inlay may be detected. Examples of the almuce shown alone occur at Arundel (Sussex), Bury St Edmunds (Suffolk), Byfleet (Surrey), Eton College (Bucks.); elsewhere, mainly small and of late date.

The **cope** is a large cloak with a hood at the back, and is the chief mark of processional vestments. There are three kinds of cope: (1) completely plain, called a choir cope, (2) plain cope with embroidered orphreys, (3) embroidered all over. This is the rarest, though a choir cope is not often found.

The **morse** is the fastening of the

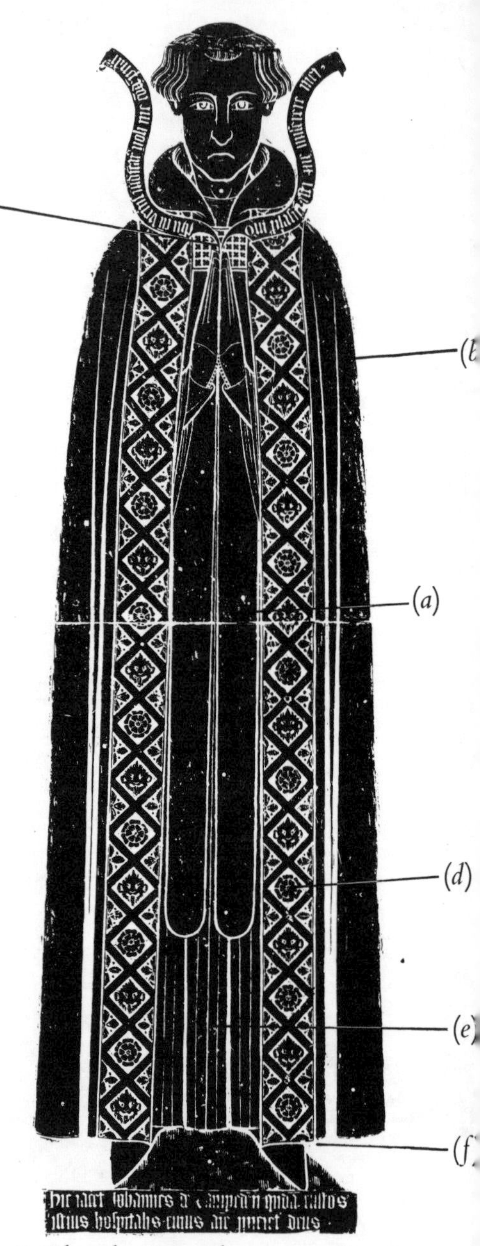

18. John de Campeden 1382, St Cross, Winchester, Hants.

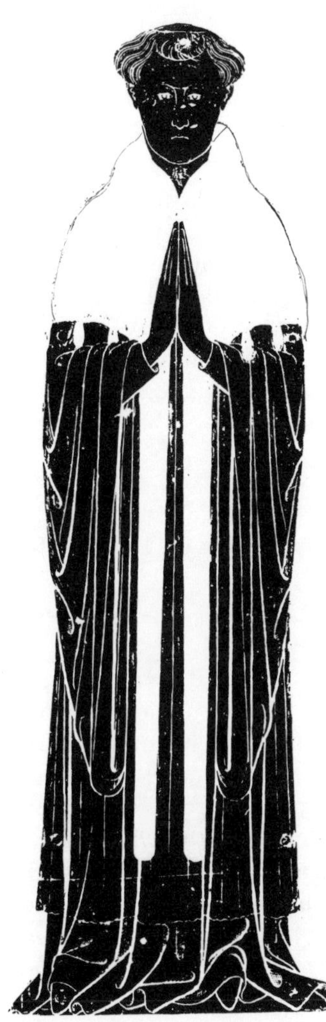

19. William Whyte 1419, Arundel, Sussex

cope, and is decorated in many ways. An initial, the letters IHC, a floral pattern, an invocation and a representation of the Trinity are among the many designs shown.

The **orphreys** are two strips of beautifully embroidered material sewn down either side of the cope. They have varied forms of decoration. Initials or floral patterns seem the most common. At Broadwater, Sussex, the brass to John Mapleton has his initial 'M' and a maple leaf, a pun on his name. At Warbleton, Sussex, a text from Job is engraved on the orphrey, the word 'credo' appearing in the morse.

The **surplice** is a loose garment worn over the cassock, only the hanging sleeves of which show.

The **cassock** is a garment like an alb but without decoration, and little of it shows beneath the cope.

* * *

Saints appear in the orphreys of many of the finest ecclesiastical brasses. At Tattershall, Lincs. and Trinity Hall, Cambridge, the twelve apostles are represented on the cope, and the same arrangement sometimes appears in canopy shafts.

The saints can be identified by the symbols which they bear. The most common ones are:

St Paul: a sword.
St Peter: two keys.
(These two saints usually appear together.)
St Andrew: a saltire cross.
St Bartholomew: a butcher's flaying knife.
St John the Evangelist: chalice and serpent.
St Catherine: a wheel.
St Margaret: a spear and dragon.
St Christopher: shown crossing a river, carrying Jesus on his back.
St George: in armour, fighting a dragon.
St Thomas of Canterbury: in archbishop's vestments.
St James of Compostella: pilgrim's staff and shell.

20. Henry Sever 1471, Merton College, Oxford

Good examples of coped priests, without saints, occur at St Cross, Winchester, Cottingham (Yorks.), Exeter Cathedral, Havant (Hants.) and Broadwater (Sussex). These are very frequent, and occur until the time of the Reformation. Those with saints are generally very large and canopied. Examples occur at Balsham (Cambs.), Castle Ashby (Northants.), Bottesford (Leics.), Ringwood (Hants.), Merton College (Oxford), and Tattershall (Lincs.).

MONASTIC COSTUME

During the Middle Ages, a great many memorial brasses were laid down to the men and women in religious orders. The monasteries were extremely rich and powerful, and these brasses were among the finest in England. With the Dissolution of the Monasteries under Henry VIII, and the religious struggles of the 17th century, most of these brasses were destroyed. From all the monastic brasses, less than thirty now remain. There are seven monks, mainly in St Albans Abbey, Hertfordshire, and four nuns in different parts of England. Of the higher offices, two priors remain. These are at St Laurence (Norwich), and Cowfold (Sussex). The last-named brass is one of the finest in England. The prior is shown wearing the habit of a Benedictine monk, his order, the Cluniac, being a branch of the Benedictines. He was prior of Lewes, the head Cluniac

21. Thomas Nelond, prior of Lewes, 1433, Cowfold, Sussex

23. Richard Billingford D.D. 1442, St Benets, Cambridge

22. Eleanor de Bohun 1399, Westminster Abbey

monastery in England, and therefore held a very important position. At Nether Wallop (Hants.) is the only surviving brass of a prioress. She wears a habit similar to that of a nun.

When their husbands died, widows would often take religious vows and join an order. These are called vowesses, and a few are seen in brass. The finest of these is Eleanor de Bohun, Duchess of Gloucester, at Westminster Abbey. Her costume is that of a nun, with the 'barbe' or chin-cloth (actually French for 'beard') over her chin.

Some of the more important brasses showing monastic costume are at Cowfold (Sussex), Denham (Bucks.), Dorchester (Oxon.), Elstow (Beds.), Nether Wallop (Hants.), St Laurence, Norwich (Norfolk) and St Alban's Abbey (Herts.). Vowesses may be found at Frenze and Witton (both in Norfolk), Quinton (Glos.), and at Westminster Abbey.

24. Sir Robert de Setvans 1306, Chartham, Kent

ACADEMIC COSTUME

There are about seventy-five examples of academic costume, mostly at Oxford and Cambridge. Since in medieval times the monks were the only general teachers, the universities were connected with the Church. Most scholars, in brass, have the tonsure, the sign of Holy Orders.

The main distinction between the robes worn by holders of different degrees of learning was the colour, and therefore on brass it is often hard to tell exactly what a man's degree is. The usual gown was called the 'cappa clausa', and was slit up the front to show the hands. This is usually worn by Doctors of Divinity. There are other types of dress, resembling the cassock of a priest but shorter. On the head was worn either a skull cap or a raised cap. Dr Billington, at St Benets, Cambridge, is kneeling, so that his hood down his back is visible. It is inlaid with a silver metal, lined to represent material.

Good collections of priests in academic dress can be found at Oxford (Magdalen College, Merton College, New College, Queen's College) and at Cambridge (King's College, Trinity Hall). The brasses in Cambridge suffered greatly under the Puritans, and only a small number of those originally laid down still remain.

ARMOUR

The first type of armour, depicted in brasses, begins in about 1260, after the Crusades. The armour consisted of a hawberk, a complete set of chain mail covering the body and the arms, a coif or hood of mail, chausses or stockings, and mail gloves. The knees were protected by metal plates, and a linen surcoat was worn over the whole armour. The brass of Sir John Dauber-

25. Sir John and Alyne Creke 1325, Westley Waterless, Cambs.

noun, Stoke d'Abernon (Surrey), 1277, shows him holding a lance and banner. He has a large sword, shield and spurs, and his feet rest on a lion. Next in date comes Sir Roger de Trumpington, Trumpington (Cambs.). His feet rest on a dog, and on his shield and scabbard are trumpets, referring to his name. His head rests on a tilting-helm and, like the very fine brass to Sir Robert de Bures, Acton (Suffolk), his legs are crossed. At Chartham (Kent) Sir Robert de Setvans is shown with his hands and head bare, and his hair curls to his shoulders. There are several theories about the significance of a knight having crossed legs and a lion or a dog at his feet, but there is little substantial evidence to prove any of these.

There are two transitional examples at Pebmarsh (Essex) and Gorleston (Norfolk), about 1320, leading up to the next distinct type. During this time several protective metal plates were added, and a few other changes made. The mail hood gave way to a helmet or bascinet, and the surcoat to the cyclas, a shorter version split up the sides, and longer at the back than at the front. Examples occur at Westley Waterless (Cambs.), Stoke d'Abernon (Surrey) and Minster (Kent).

The next type is very different from the previous one. Legs and arms are completely covered by steel plates, with the exception of the arm-pits, where the hawberk is visible. A tunic or jupon, often bearing heraldic charges, is now worn over the suit of mail. Steel gauntlets are now used instead of mail gloves. The helmet has a mail tippet attached, covering the head and shoulders. The sword becomes shorter, and is hung at the side now, and daggers are shown. There are numerous examples of this type of armour, the series at Cobham (Kent) being particularly fine.

26. Sir John and Joan de la Pole 1380, Chrishall, Essex

The next type, used mainly during

27. Sir John Lysle 1425, Thruxton, Hants.

28. Sir William and Margaret Vernon 1467, Tong, Salop

29. Sir Thomas Bullen K.G. 1538, Hever, Kent

the period of the Lancastrians, is completely composed of steel plates. A new device came into use, the gorget, a steel circle round the neck to replace the tippet. The helmet became domed instead of pointed, and the elbows were protected by heart-shaped plates, giving way later to fan-shaped ones. The armour of this time was heavy but bearable. As with ladies' costume at this time, armour became very stylized, and many brasses are almost identical.

The most noticeable feature of armour around the time of the Yorkist kings is the great size of the plates, particularly at the elbows. The helmet is rarely depicted on brasses now, but a helm is sometimes shown behind the head. The sword is hung across the body once more, but does not regain the great size of the previous period. The armour is incredibly heavy, and it is remarkable that knights were able even to move in it. During this period, the engraving of brasses deteriorated, like the making of English armour. At Hildersham (Cambs.) there is a good canopied knight; other good examples occur at Isleham (Cambs.), Morley (Derbys.), Sherbourne (Norfolk), Thornton (Bucks.) and Tong (Salop.).

Just before the Tudors, heraldic tabards become popular. Though earlier examples exist, it is not till now that they are found in profusion. Although still used, armour begins to die out. The sword returns to its position by the side, and the foot-armour becomes round instead of pointed. Helmets are totally abandoned; knights are shown with shoulder-length hair. The brass to William

Here lyeth Interred the body of
Nicholas Wadham whiles he lyved
of Merefeild in ye County of Somer-
sett Esquier, ffounder of Wadham
Colledge in Oxforde, who Depted
this lyfe ye xx day of Octob 1609

Here lieth also ye body of Dorothie
Wadham widow, late the wife of
Nicholas Wadham Esqr; Foundresse
of Wadham Colledge in Oxforde
who died the 16 of May 1618
In the yeare of her age 84.

30. Nicholas and Dorothy Wadham
1618, Ilminster, Som.

31. Lady Margarete de Camoys 1310, Trotton, Sussex

Catesby at Ashby St Ledgers (Northants.) is a particularly good canopied example (*see* page 18).

Sir Thomas Bullen, Hever (Kent), the father of Anne Boleyn, wears the full insignia of the Order of the Garter. Beneath his robes his armour can be clearly seen. About six brasses show the Garter tied below the left knee, Trotton (Sussex) being a good example. Canons of Windsor wore the garter badge on their copes, as at Eton College Chapel.

During the reigns of Queen Elizabeth and the Stuart kings, armour was used almost solely for decoration. This armour is intricate and beautiful. The breastplate became of peascod form, that is raised down the centre, and long tassets covered the thighs. A small ruff encircles the neck, and a helm is sometimes shown behind the head. At West Firle (Sussex) there is a series of brasses to the Gage family, designed and engraved by Gerard Johnson in 1595. These brasses are small and very well executed, and the original drawings for them still survive at Firle Place. The Ilminster (Somerset) brasses, to the founders of Wadham College at Oxford, are beautifully designed and very impressive, paying great attention to minute details (*see* pages 21–2).

With the general use of firearms, from the Civil War onwards, armour ceased to be used, and is no longer depicted in brasses.

LADIES' COSTUME

The earliest brasses of a lady depict a life-size figure in a kirtle with tightly-buttoned sleeves, over which was worn a flowing gown. In some cases a corded mantle is shown. A wimple covers the chin, and a veil the head, showing a fillet keeping the hair in

32. Lady Elizabeth Cobham 1375, Lingfield, Surrey

33. Margaret Cheyne 1419, Hever, Kent

place. The two earliest brasses of this sort are Lady Camoys, Trotton (Sussex), about 1310, and Joan Kobham, Cobham (Kent), about 1320. The dress of Lady Camoys originally contained nine enamel shields, which have since been stolen.

The next style of dress was a low kirtle, usually covered by a cote-hardi, a garment similar to the kirtle but with sleeves terminating at the elbow and long hanging lappets. Another dress in vogue at this time was the sideless cote-hardi, a skirt supported by two Y-shaped strips of fur joined at the shoulders. A mantle was often worn over this type of dress. The hair is worn in the nebulé

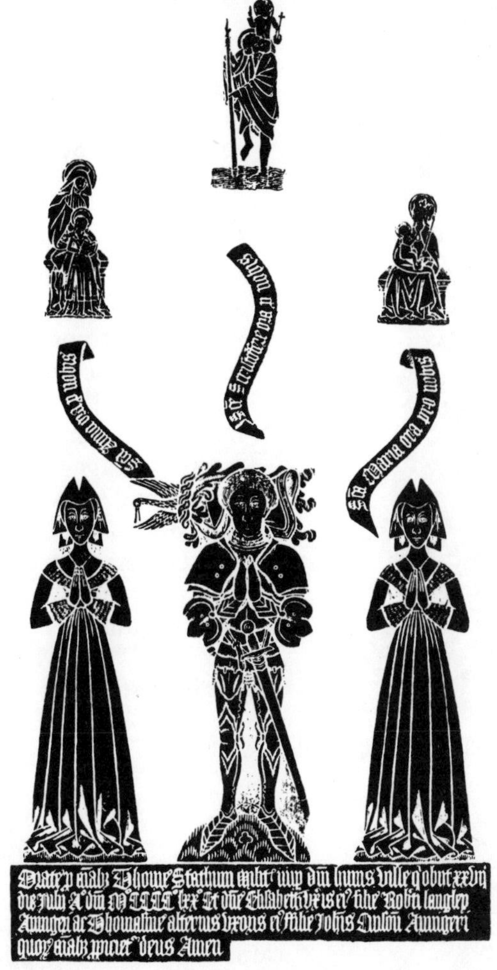

34. Sir Thomas, Elizabeth and Thomasine Stathum 1470, Morley, Derbys.

or zigzag style, and the head is often shown resting on an elaborate cushion.

During the reign of the Lancastrian kings, a stylized type of costume appears. The waist is extremely high and thin, and a long mantle is worn. The space between the dress and the mantle is sometimes inlaid with lead, to represent fur. The hair is enclosed in square nets on either side of the head, and a short veil is placed over the nets. Often the hair is placed in a horned arrangement on the head, as at Thorncombe (Dorset). (*See* page 43.)

This style leads up to the mitred head-dress in the Yorkist period. With this comes a new type of dress, flowing, with a high waist, and low, often furred, collars. These brasses are frequently small, and there are numerous examples of men with two, three or even four wives, all dressed in this costume. A good instance is at Thornton (Bucks.), where Robert Ingylton stands with his three wives beneath a quadruple canopy. At Morley (Derby),

35. Margaret Peyton 1484, Isleham, Cambs.

Sir Thomas Strathum and his two wives, Elizabeth and Thomasine, pray to St Anne, St Christopher and the Virgin Mary.

In the reign of Edward IV came a very strange head-dress indeed. It is called the 'Butterfly'. It is a veil arranged over a complicated set of wires, but is not shown to advantage on brasses since it appears cumbersome and heavy. Jewellery in the form of heavy rings and vast elaborate necklaces now appears, and also heraldic dresses and mantles become popular. The brass to Lady Say, Broxbourne (Herts.) retains its original red enamel. (*See* page 16.) Margaret Peyton, Isleham (Cambs.) wears an embroidered dress, and 'Jesu Merci' is engraved on her butterfly head-dress.

36. Elizabeth Perepoynt 1543, West Malling, Kent

37. Mrs Ann Kenwellmersh and grandson, Meneleb Rainsford, 1633, Henfield, Sussex

38. Sir Edward and Elizabeth Gage 1595, West Firle, Sussex

The pedimental head-dress began in the reign of Henry VII. It is heavy, and completely hides the hair. The corresponding dress also was of heavy material, with tight sleeves and a square-cut neck. Often a pomander or scent-box was hung from the belt. Most of these brasses seem clumsy, compared with the graceful appearance of earlier styles.

The next head-dress is the 'Mary Queen of Scots' cap, otherwise known as the French bonnet. Dresses changed to the farthingale and stomacher, and a small ruff was worn round the neck. This costume is relatively simple, and there are several good examples, such as the Gage brasses at West Firle (Sussex).

The final period has few examples in brasses. The ruff grew to a tremendous size, and a large flowing veil now appears. The dress is very ornate, with lots of frills, bows and embroidery. It is in some ways similar to the farthingale, but without the plain stiffness of previous times. Shoes, which for centuries have been small or flat, now become high-heeled, a modern touch, from about 1630 onwards.

CIVILIAN COSTUME

The first male civilian brasses appear about the time of Chaucer, with the rising middle class of the second half of the 14th century. The dress is usually a long robe, buttoned at the neck, and a hood. This can be seen at Great Berkhampstead (Herts.), Ore (Sussex) and St Alban's, St Michael (Herts.). Another type of costume is a tunic reaching below the knees, with a slit up the front and long lappets hanging from the sleeves. This can be seen at Taplow (Bucks.) and King's Lynn. Civilians have long hair and beards until about 1410.

In the reign of Richard II, a long mantle buttoned on the right shoulder was worn. An anelace, or short sword, hung from the belt of the tunic. This is well illustrated by the brass to a yeoman and his brother, a priest, at Shottesbrooke (Berks.). In the brass at Hildersham (Cambs.) the mantle is thrown back to show the tunic, hose and anelace underneath (*see* page 55).

Next comes the style most frequently shown on brasses. A long tunic is worn with very wide sleeves, caught tight at the wrists. These sleeves have been called 'Devil's pockets', as stolen goods could easily be concealed in them. At Stopham

39. Civilian and brother, a priest, 1370, Shottesbrooke, Berks.

40. Nichole de Aumberdene 1350, Taplow, Bucks.

41. Sir Thomas and Joan Brook 1437, Thorncombe, Dorset

42. Henry and Joan Hatche 1533, Faversham, Kent

43. John Bartellot 1630, Stopham, Sussex

(Sussex) Richard Bertlot wears this costume with a livery collar, and carries his staff of office as Marshal of the Hall to the Earl of Arundel. The brass at Thorncombe (Dorset) shows this costume particularly well. Hair is now worn short, and above the ears. This was quite possibly due to the fact that hair could easily get caught in the new type of armour.

In the time of the Tudors, a new style of dress became fashionable. It was a long furred gown, with quite full hanging sleeves trimmed with fur. The collar was also fur-trimmed, and often a bag and a rosary hang from the belt. With the dying out of armour, hair is worn shoulder-length, and shoes are round-toed instead of pointed.

There are numerous examples, almost all extremely poor brasses, of the familiar doublet and hose, covered by a long cloak; a small ruff was worn round the neck. A late example at Stopham (Sussex), *c.* 1630, shows the high jackboots and lace collar of the Stuart kings. The last figure brass, 1773, at St Mary Cray (Kent) is very poor.

There was a revival of monumental brass engraving during Victorian times, but these were mainly copies of medieval brasses. Today some clergy are commemorated by brasses showing them in academic costume or the rather similar post-Reformation costume – this latter costume is well illustrated at Battle (Sussex), 1615. As yet no one has attempted to make a brass in modern civilian costume.

44. Woolmerchant and wife 1400, Northleach, Glos.

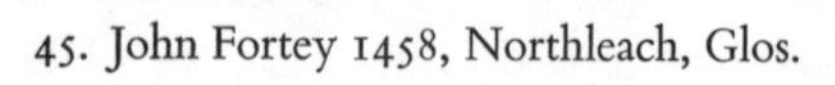
45. John Fortey 1458, Northleach, Glos.

WOOL MERCHANTS

In the Middle Ages, England's main source of wealth was her wool trade. Large flocks of sheep roamed all over the Cotswolds, which was one of the chief centres of the trade. These flocks were owned by very rich men, who were members of the Staple of Calais, a great council of wool merchants which regulated the export of wool and woollen cloth.

Such men built huge churches and gave large sums of money to the Church. John Fortey built the beautiful perpendicular nave of Northleach church, in which he lies buried. Beneath his feet are a woolsack and a sheep, denoting his trade. In the marginal inscription are six roundels, containing his mark and his initials. This mark was stamped on all the sacks of wool produced by his sheep. His canopy is in an unusual style, and though it has been mutilated, it still gives the light effect of perpendicular architecture. There are several more brasses to wool merchants at Northleach, and at other churches all over the great wool-producing areas such as Chipping Campden and Cirencester, which, like Northleach, are in Gloucestershire, and Linwood and Stamford in Lincolnshire.

JUDGES AND LEGAL COSTUME

A most important section of medieval society was the legal profession. There are several brasses to judges, of which three are outstanding, those at Brightwell Baldwin (Oxon.), Deerhurst (Glos.) and Graveney (Kent). On each of these brasses the judge wears a coif, or close-fitting cap, a hood, and a fur-lined mantle, buttoned on the right shoulder. Each has a fine canopy, the Deerhurst brass showing St Anne and formerly St John the Baptist, one on each side of the central pinnacle. The judge's wife has a small dog at her feet, with the label 'Terri'. This is the only remaining instance of a named pet in brass. The Graveney brass is about ten feet high. The judge and his wife each hold a small heart. At Brightwell Baldwin, John Cottusmore and his wife are shown on two different brasses. In the first they are very small, kneeling to a bracket which is now lost. In the second they stand beneath a fine canopy, and their children kneel at their feet.

Other examples of judges occur at Bray (Berks.), Cople (Beds.), Dagenham (Essex), Gunby, near Spilsby (Lincs.), Latton (Essex), Rougham (Norfolk), St Mary Redcliffe, Bristol, Watford (Herts.), and elsewhere.

46. Notary 1475, St Mary Tower, Ipswich, Suffolk

47. Sir John and Alice Cassy 1400, Deerhurst, Glos.

48. John Rede 1404, Checkendon, Oxon.

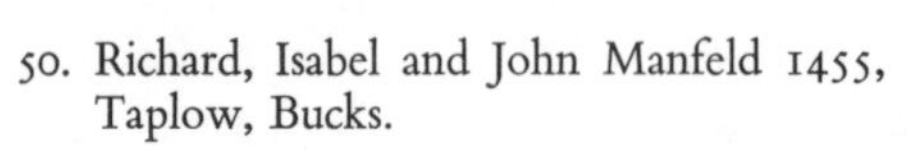

50. Richard, Isabel and John Manfeld 1455, Taplow, Bucks.

49. John Corp and grand-daughter, Eleanor, 1391, Stoke-Fleming, Devon

Among other legal brasses are those of notaries. At St Mary Tower, Ipswich, there is a brass of a notary with an inkhorn and pencase suspended from his belt, and a small hat on his left shoulder. There are also a few examples of Sergeants-at-law, the best being John Rede at Checkendon (Oxon.). He wears a plain gown and a hood. This brass is particularly interesting, since his left arm is rather longer than his right. This may be an error in engraving, or an actual likeness of John Rede.

CHILDREN

Children are often shown in brasses at the feet of their parents, as at Brightwell Baldwin (Oxon.) and at Cirencester (Glos.). In the 16th and 17th centuries, when rectangular-shaped plates became popular, they are shown kneeling, the girls behind their mother and the boys behind their father. Children who died young are shown either as swaddled chrysoms or in shrouds, while children alive when the brass was laid down are in costume. Their clothes are the same as those of their parents and unmarried girls usually have their hair loose. The brass at Taplow (Bucks.) shows the children of Robert Mansfield; these are Richard, who died aged nineteen, Isabel, who also died fairly young, and 'yong John', who had died before his brother and sister.

At Stoke Fleming (Devon), there is another unusual brass to John Corp and his grand-daughter Eleanor, who died in 1391. To bring the figures to about the same height, Eleanor has been placed on a pedestal; and incidentally, as John was a ship-man, the canopy is based on a ship's poop. One of the latest brasses to show a child is that of Mrs Ann Kenwellmersh and her grandson Meneleb Rainsford, dated 1633, at Henfield (Sussex). He wears the costume shown in many paintings by Van Dyke, and holds a small round hat. (*See* page 40.)

CANOPIES

Canopies are found over some of the largest and most splendid brasses.

There are three main types of canopy: single, double, and triple. These terms refer to the number of arches in the canopy. Saints often appear in canopy shafts, and can be identified by their symbols (*see* p. 27).

Canopies need to be rubbed carefully, because they are usually very intricate; it is very easy to leave out a piece of foliage or the top of a pinnacle. Through the centuries, most canopies have lost pieces; it is necessary to check very thoroughly before you begin rubbing, and even to make a sketch diagram to show missing parts.

Good examples of canopies are:

Single: Higham Ferrers (Northants.), Horley (Surrey), Hurstmonceaux and Warbleton (Sussex) and at Westminster Abbey.

Double: Dartford (Kent), Deerhurst (Glos.), Great Tew (Oxon.), Linwood (Lincs.) and Trotton (Sussex).

Triple: Balsham (Cambs.), Bottes-

51. Baron Thomas and Elizabeth Camoys 1421, Trotton, Sussex

52. John Sleford 1401, Balsham, Cambs.

ford (Leics), Cobham (Kent), Cowfold (Sussex), New College, Oxford, Northleach (Glos.), Thruxton (Hants.) and at Westminster Abbey.

CROSSES AND BRACKETS

A great number of crosses were laid down before the Reformation, but owing to their destruction by Thomas Cromwell's men, few now remain. Those few do, however, show us how very beautiful this form of memorial was. Crosses fall into two main groups, plain crosses and those with figures.

The plain crosses rise from a base of steps, and are very simple. The arms of the cross usually terminate in fleur-de-lis. At Higham Ferrers (Northants.) the arms terminate in the four symbols of the Evangelists: the angel of St Matthew, the lion of St Mark, the ox of St Luke and the eagle of St John. Examples of plain crosses also occur at Beddington (Surrey), Broadwater (Sussex) which is inscribed, Cassington (Oxon.), and Hever and Penshurst (both in Kent), the last two being without fleur-de-lis.

The second type of cross, that with figures, itself falls into two categories: those where the deceased appears in the cross, and those where the deceased kneels at the foot of a cross containing a saint. Of the first, there are quite a number of examples, usually mutilated. At Taplow (Bucks.) the cross to Nicholas Aumberdene, a fishmonger, rises not from steps but from a fish's back. There are several

53. Robert de Waldeby, Archbishop of Dublin, 1396, Westminster Abbey

55. Robert and Eleanor Parys 1408, Hildersham, Cambs.

54. Britellus Avenel 1408, Buxted, Sussex

56. John Strete 1405, Upper Hardres, Kent

examples of crosses in Merton College, Oxford, and in Kent at East Wickham, Stone and Woodchurch. The cross to Britellus Avenel at Buxted (Sussex) is remarkably complete.

There are many indents remaining to show that it was a popular arrangement to have a saint in the cross and the deceased kneeling at the foot. The two best surviving brasses are at Newton-by-Geddington (Northants.), where John Mulsho and his wife kneel to St Faith, and at Hildersham (Cambs.), where Robert Parys and his wife kneel to the Holy Trinity.

Bracket brasses are in many ways similar to crosses. A long shaft supports the bracket, on which the figures stand. A canopy is often placed over the figures. Examples occur at Bray (Berks.), Cobham and Southfleet (both in Kent), and Merton College, Oxford. At Upper Hardres (Kent) there is a unique arrangement, where John Strete kneels at the foot of a bracket supporting St Peter and St Paul.

SKELETONS AND SHROUDS

Skeleton and shroud brasses are usually found in the 16th and 17th centuries. The skeleton brasses are often very grim. Usually a skeleton appears in a knotted shroud. The 'shroud' brasses show a living figure in a shroud. These are often very beautiful, as the women have long hair

flowing down outside the shroud. As in later stone monuments, children who died young are sometimes represented by skulls, and babies are shown wrapped in swaddling clothes.

Skeleton brasses occur at Margate (Kent), Norwich, Corpus Christi, Oxford, and Hildersham (Cambs.). At Oddington, near Oxford, Ralph Hamsterley is shown as being eaten by worms.

Shrouded figures occur at West Firle (Sussex), Digswell (Herts.), Childrey (Berks.), where they rise from coffins, and Yoxford (Suffolk).

Children in swaddling clothes occur at Long Melford (Suffolk), Stoke d'Abernon (Surrey) and Marden (Herefs.).

57. John Bloxham and John Whytton 1420, Merton College, Oxford

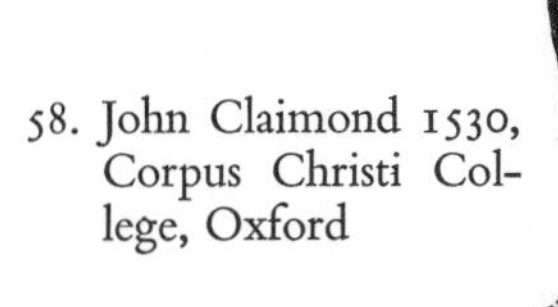

58. John Claimond 1530, Corpus Christi College, Oxford

HEARTS AND CHALICES

'Heart' brasses are quite rare. Usually the heart is held by a pair of hands issuing from a cloud. Occasionally it is inscribed or has drops of blood on it. Heart brasses sometimes commemorate the burial of the heart alone, but there is not enough documentary evidence to tell whether this is a general rule or not. Sometimes the heart is held in the hands of the deceased or shown over a figure. Usually a heart by itself commemorates a priest.

In Yorkshire and Norfolk there are a number of 'chalice' brasses. These commemorate priests, and were probably laid down if there was not enough money to pay for a figure brass. The chalice and wafer appear, sometimes held by a priest, usually in the 16th or late 15th centuries. Some of the finest early ecclesiastics also have a chalice, either between their hands or lying on their breast.

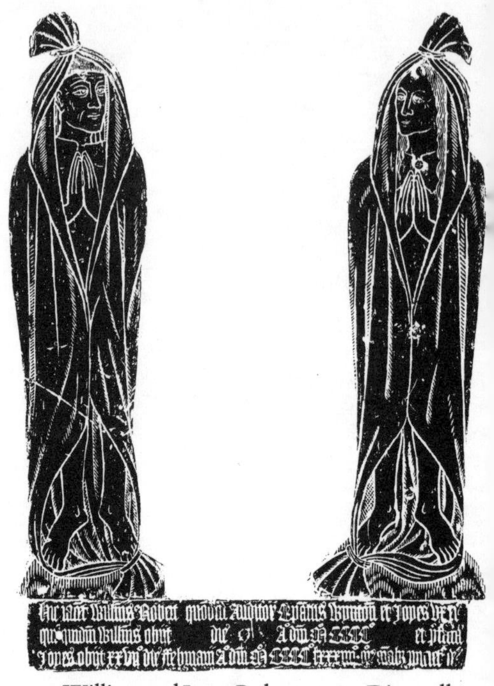

59. William and Joyes Roberts 1484, Digswell, Herts.

Examples of heart brasses occur at Caversfield and Lillingstone Lovell (Oxon.) and at Fakenham, Helhoughton, Ludham, Southacre and Trunch (all in Norfolk). Chalice brasses occur at Ripley and Leeds in Yorkshire, Gazeley and Rendham in Suffolk, and numerous places in Norfolk, such as Buxton, Colney, Guestwick, North Walsham and Scottow. The brass at Holwell (Herts.) is most strange, and shows a chalice below two woodiwoses or wild men, a pun on the name Robert Wodehowse. Many such rebuses, or pictorial puns, are found in brasses. Sir Roger de Trumpington (*see* outside cover of book) displays a trumpet on his shield.

60. Anne Chute and sister, Frances, in swaddling clothes 1614, Marden, Herefordshire

61. Chalice and wild men to Robert Wodehouse 1510, Holwell, Herts.

62. Heart to Thomas Denton 1533, Caversfield, Oxon.

SHIELDS AND COATS OF ARMS

The study of Heraldry on brasses is of great importance. Often there are traces of original coloured enamels left, so that the shield can be rubbed in its correct colours. With a little knowledge of simple heraldry one can find out quite a lot about people commemorated on brasses.

The following is a list of heraldic colours, fur and metals (using English names) found most frequently on brass shields.

Colours	*Fur*	*Metals*
Black	Ermine	Silver
Red		Gold
Blue		
Green		
Purple		

Except for the fur, all these can be rubbed by using different coloured waxes.

When trying to reconstruct original colours, there is one simple rule of heraldry that will prevent any mistakes being made: 'Never put a metal on a metal or a colour on a colour.'

Quite often, in Tudor times, a person's coat of arms is shown on his tabard. The brass at Hunstanton (Norfolk) is a particularly fine example. Ladies wear their own arms quartering their husbands' on their dress or mantle. Elizabeth Goringe, Burton (Sussex), is a unique example, as she is shown wearing a man's tabard. These heraldic brasses can look very fine indeed, if they are rubbed using the correct heraldic colours.

Another interesting aspect of

63. Elizabeth Goringe 1558, Burton, Sussex

heraldry is the study of footrests. Generally knights have their feet resting on a lion, though this is not always the case. Two early knights, from Trumpington (Cambs.) and Pebmarsh (Essex), have dogs at their feet, while John Cassy, a judge, from Deerhurst (Glos.) and John de Grofhurst, a priest, from Horsmonden (Kent), both stand on lions. Ladies often have pet lap-dogs in the lower folds of their dress.

Sometimes a man's feet rest on the symbol of his trade: woolsacks and sheep for a woolmerchant, scissors for a tailor and wine casks for a vintner. In later brasses the persons commemorated stand on small grass mounds or on chequered pedestals.

There are a number of brasses where a heraldic beast appears at the foot of a person, such as the unicorn of the Chaucers at Ewelme (Oxon.), the bear of the Beauchamps at Warwick, and the elephant and castle of the Beaumonts at Wivenhoe (Essex). A wide range of animals appears, from a hedgehog at Digswell (Herts.) to a large griffin at Hever in Kent (*see* page 35). At Tong in Shropshire (*see* page 34), Margaret Vernon has a strange elephant-like creature at her feet. This is probably a dragon, the symbol of St Margaret. Elephants had not been seen at large in England since the Claudian invasion in Roman times, though more recently an elephant had been given to Henry III in 1255, and was kept in the Tower of London. It is very possible that an ordinary person living in the Middle Ages would imagine an elephant as a kind of dragon. The study of footrests is fascinating as they are among the most varied details to be found in brasses.

OTHER UNUSUAL BRASSES

There are a number of interesting and unique brasses which do not fall into any of the categories previously mentioned; a few are described here.

At Edlesborough (Bucks.) there is a 'rose' brass. This brass was thought to be unique, but another rose has been located and returned to Mawgan-in-Pyder (Cornwall). Roses also quite often appear as part of the design in larger brasses. The rose was the symbol of the Virgin Mary, and recalls the medieval rhyme:

There is no rose of such vertu
As is the rose that bare Jesu,
Alleluia.

64. Roger de Strange 1506, Hunstanton, Norfolk

65. Inscribed rose 1412, Edlesborough, Bucks.

Several brasses commemorate a person's trade. At Baldock (Herts.) a huntsman has a rope, dagger and hunting horn hung from his belt. Peter Devot, a glover, from Fletching (Sussex) is commemorated by a pair of gloves. At Margate (Kent) there is a small plate engraved with a ship for a ship builder, Roger Morris. An indent at Barking (Essex) shows a large bell over the figure of a bellfounder. Some brasses show a particular event in a person's life or the way he died. At Walton-on-Thames (Surrey), John Selwyn is shown riding a stag, which he once did in front of Queen Elizabeth. Bishop Robert Wyvil, Salisbury Cathedral, is shown standing in Sherbourne Castle which he recovered for the Church. The rabbits at the foot of the castle recall his recapture of the Chase of Bere (*see* page 24).

James Gray, a park keeper, who died while shooting a stag, is shown in a brass at Hunsdon (Herts.). He is killing the stag with a crossbow, while a skeleton, representing Death, stabs him with a dart. At Llandinabo (Herefs.) there is a brass to Thomas Tompkins, a young boy who was drowned; he is shown standing in a small pond.

There is only one brass in England commemorating a king. This is to St Ethelred, king of the West Saxons, who was martyred by the Danes in 871. This small half-effigy was engraved in about 1440, and was placed over his remains in Wimborne Minster (Dorset).

66. Gloves to Peter Devot 1440, Fletching, Sussex

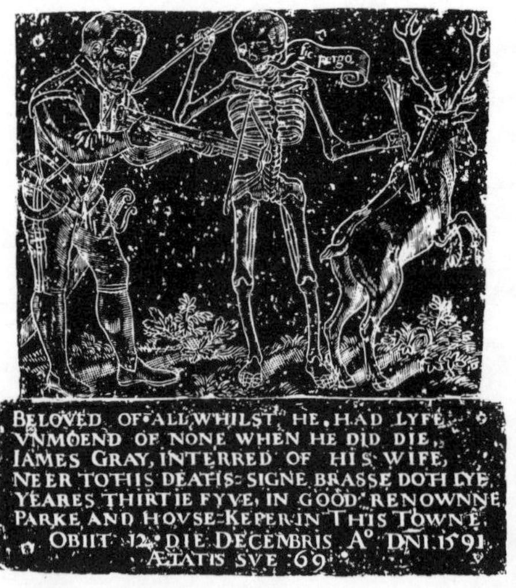

67. James Gray 1591, Hunsdon, Herts.

68. King Ethelred engraved 1440, Wimborne Minster, Dorset

69. Thomas Tompkins 1629, Llandinabo, Herefordshire

RELIGIOUS SCENES IN BRASSES

Besides these unusual brasses, there are a number of religious scenes in brass, where the local craftsman often shows remarkable inspiration. The Holy Trinity frequently appears above kneeling figures and in canopies. The Father is usually shown as an old man with a long flowing beard, and the Dove perches on one arm of the crucifix. Good examples occur at Childrey (Berks.) and also Hildersham (Cambs.) where the Dove is omitted (*see* page 55).

There are only three examples of the Annunciation: Fovant (Wilts.), March (Cambs.) and Hereford Cathedral. The Hereford annunciation is in the centre of a Renaissance canopy. The Virgin kneels at a prayer-desk, her head inclined to face the angel, who wears a long robe blown by the wind, and carries a jewelled sceptre. Between them is a double-handled vase containing a spray of lilies, and the Dove descends from the centre of the canopy in a stream of light.

The Resurrection is more widely found: examples occur at All Hallows, Barking, Cranley (Surrey), Slaugham (Sussex), Stoke Charity (Hants.) and elsewhere. There are two types of Resurrection, those with soldiers round the tomb and those without. The Stoke Charity brass is a most beautiful representation of the latter type. In the former, as at Cranley, the soldiers, mentioned in St Matthew as being sent by Pilate to guard the sepulchre, are shown in the armour used at the time of engraving. They

70. Annunciation 1500, Fovant, Wilts.

71. Nativity 1500, Cobham, Surrey

72. Resurrection 1525, Slaugham, Sussex

lie round the foot of the tomb, and Jesus is shown standing, so that three of the five wounds are visible, the feet being concealed in the tomb.

At Cobham (Surrey) there is a unique brass of the Nativity. It is like a child's drawing; the stable is squashed in behind the main figures. As in medieval paintings, these figures are much larger than the less important ones; one of the shepherds is not much larger than the baby. Mary is lying with her head on a cushion, Joseph stands beside her, and the ox and ass are peering into the manger.

These are some of the familiar religious scenes found in brasses, paralleled in stained glass and statuary. At Hereford Cathedral there is a particularly fine collection of religious brasses, including the unfamiliar scene of the beheading of John the Baptist, not found elsewhere in brass. These small brasses make very beautiful rubbings, and are well worth collecting for their artistic quality alone.

INSCRIPTIONS

The reading of inscriptions is a difficult task, and can only be done with practice. One has to be able not only to translate them from the Latin, or in earlier brasses from Norman French, but also, in the first place, to read the lettering. This table shows the approximate dates at which each form of lettering and language was used:

1250–1350	Lombardic script	Norman French
1350–1500	Black letter with contractions	Latin
1500–1650	Tudor black letter or capitals	English

There are, of course, exceptions to this. At Brightwell Baldwin (Oxon.) there is an English inscription to 'John ye Smith', who died about 1370. Many brasses to clergy had Latin inscriptions after the Reformation, as, although church services were now in English, Latin was still the language of the educated.

Norman French inscriptions are not usually too hard to read. They are mostly short, merely giving the name of the person commemorated, and a brief prayer at the end. Anyone with a knowledge of simple French and good guesswork can usually decipher early inscriptions, such as this from Stoke d'Abernon (Surrey):

Sire: John: Daubernoun: Chivalier: Gist: Ici: Deu: De: Sa: Alme: Eyt: Mercy.

Note that 'gist' would now be 'gît'.

On these very early brasses, the inscription is set in simple letters round the edge of the stone or on a small plate beneath the figure.

The Latin inscriptions in black letter are often exceedingly difficult to decipher. Vowels are usually left out, and a bar appears above the previous letter or somewhere near the omission. The date is almost always in Roman numerals, as arabic numerals were very little used at this time. Most inscriptions begin 'Hic jacet' – here lies – and end with an abbreviated form of prayer 'on whose soul(s) may

73. Sir John Wantele 1424, Amberley, Sussex

God have mercy'. Inscriptions in the larger brasses of this time are set on a continuous strip of metal, running round the brass like a frame. The start, which is usually at the top left-hand corner, is marked by a cross. The following are a very few of the words and expressions that one is likely to meet; but I suggest that a Latin dictionary is at hand when trying to decipher these types of inscription: miles ('knight'), armiger ('esquire'), eccles (abbreviation for 'church'), hic jacet ('here lies'), orate pro anima ('pray for the soul of'), cui' aĩe ppiciet' de' amẽ (abbreviation for 'on whose soul may God have mercy').

The English inscriptions are much less hard to read. The spelling is generally erratic, and will differ from brass to brass, as no uniform dictionary or standard form of spelling was made until the 18th century. Often these inscriptions rhyme, usually clumsily and in irregular lines. Some verses are touching and beautiful, such as these to a small boy at Henfield (Sussex) (*see* page 40).

Great Jove hath lost his Ganymede I know
Which made him seek an other here below
And findinge none, not one, like unto this
Hath ta'ne him hence into eternall bliss
Cease then for thy deer Meneleb to weep
God's darlinge was too good for thee to keep
But rather joye in this great favour given
A child on earth is made a Saint in Heaven

BRASSES OF FOREIGN WORKMANSHIP

Several brasses still existing were imported to Britain from the Continent. These have a different appearance from those of English workmanship. They are usually very large rectangular plates, with a figure or figures standing against a background of flowers or beasts. Many also show elaborate canopies, often filled with saints, angels and mourners. These brasses were engraved on the Continent, then transported to England, with the stone matrix, and the brass was subsequently laid down. Old documents remain giving orders for foreign brasses and the prices of manufacture. Between one-third and half the total cost was for transporting the engraved brass to England.

The brass at North Mimms (Herts.) was, when engraved, a rectangular plate. The English workmen, who prepared the stone slab, cut away the floral background to make it like the English type of brass. The brass of Sir Simon de Wensley, at Wensley (Yorks.) shows just the figure of a priest, but life-size, and engraved in such a style, with the eyes closed and the chalice on his breast, that it is undoubtedly of foreign workmanship. This is a list of the chief foreign brasses in England: All Hallows, Barking (Essex), Elsing (Norfolk), Christ Church Mansion Museum, Ipswich (Suffolk), King's Lynn (Norfolk), Newark (Notts.), All Saints, Newcastle (Northumberland), North Mimms (Herts.), St Albans

74. Priest 1380, North Mimms, Herts.

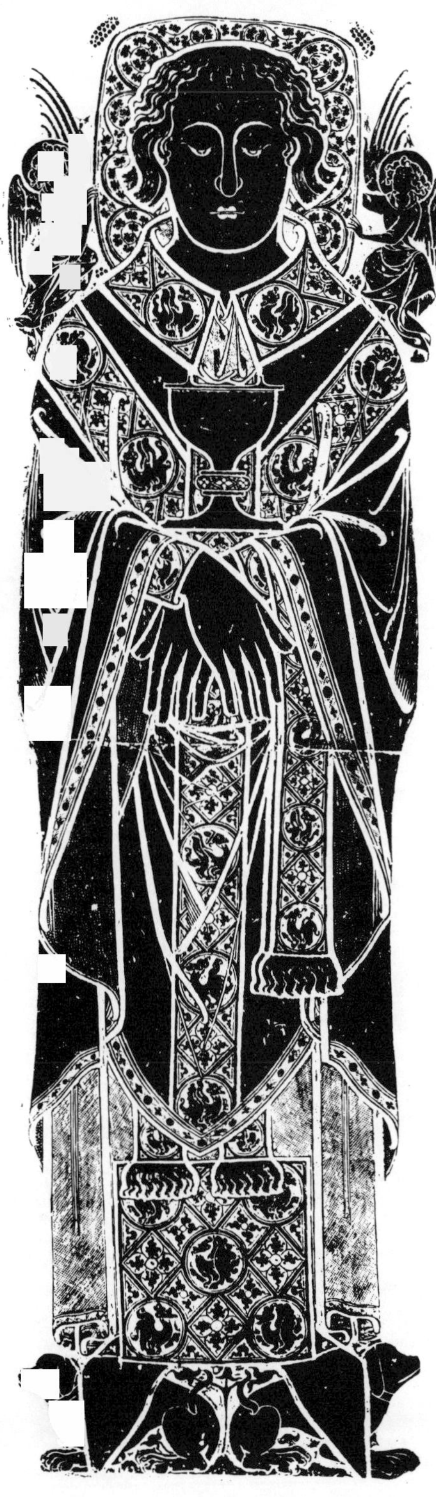

Abbey (Herts.), Topcliffe and Wensley (both in Yorkshire), and Fulham (Middlesex).

There are other examples which are possibly of foreign workmanship or influence, but which are not as splendid as these.

There are still some brasses left on the Continent, but few. The fury of the French Revolution destroyed nearly all the most beautiful French brasses. There are still a number in Germany, and these bear strong resemblances to their English counterparts, some of them probably coming from the same schools of engraving.

Brasses of foreign workmanship are often very large indeed, and need paper wider than one can generally buy; though it is possible to buy 80-inch paper (*see* p. 10), this is very difficult to transport. They, therefore, have usually to be rubbed in strips, then carefully fastened together on the back. It is hard to get the join accurate, and great care has to be taken in doing this. Sometimes foreign brasses have been mounted on walls, such as one of the two large brasses at King's Lynn. It is often difficult to get Sellotape to stick to a wall and support the weight of paper needed to cover such a brass. One method is to stick the paper firmly to the top edge of the brass – Sellotape will hold much more firmly on a metallic surface – then rub the top on a separate strip of paper, and join it on later.

75. Sir Simon de Wensley 1375, Wensley, Yorks.

76. Thomas and Emme Pownder 1525, Christ Church Mansion Museum, Ipswich, Suffolk

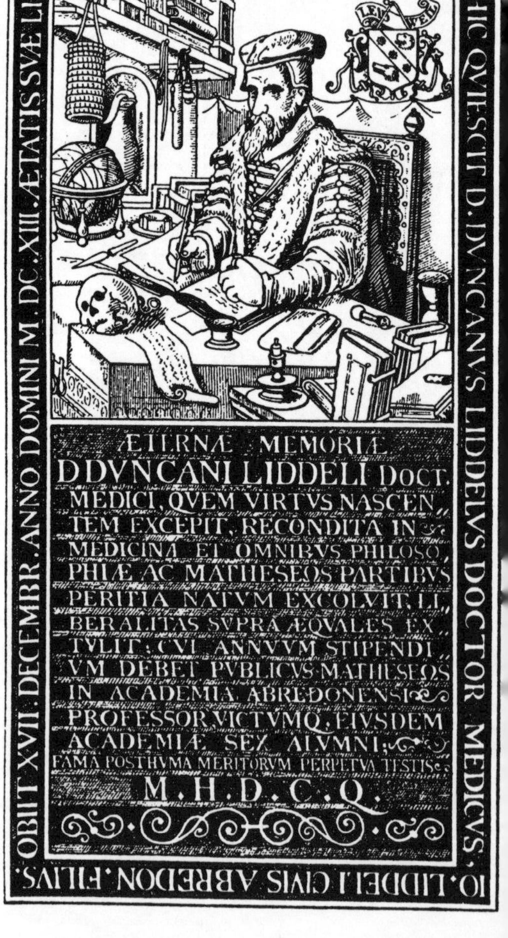

77. Duncan Liddel M.D. 1613, St Nicholas, Aberdeen

Part Three: HOW TO FIND AND IDENTIFY ALL FIGURE BRASSES

Brasses are widely distributed throughout England. East Anglia and Kent are particularly rich in brasses, while there are fewer in the West of England. About half are figure brasses.

The following lists, based on the list compiled by Mill Stephenson, include all the figure brasses in the British Isles. They are arranged by counties, and give a brief description of each figure brass, i.e. type, date of death and details such as canopies. Also included are the rarer types of brasses, hearts, chalices and other symbols of interest. Some indication of size is given, and it may be assumed that canopied brasses are usually quite large, about 8 feet.

The conditions for brass-rubbing, of course, can change if a new incumbent is appointed, and it should be remembered that permission is not invariably granted. To rub at places like Westminster Abbey, London, Stoke d'Abernon (Surrey) and Cobham (Kent), it is often necessary to book up to six months in advance as the waiting lists are very long. At most places permission is easy to obtain, and you can find and identify the brasses you wish to rub in the following pages.

List of all the Figure Brasses in Churches throughout the British Isles

Key

A = armed man.
ac = academical dress.
B. V. = Blessed Virgin Mary
C = man in civilian dress.
can = canopy.
demi = half-effigy.
E = priest in either mass or processional vestments.
fgn = of foreign workmanship.
her = in heraldic dress.
kng = kneeling.
L = a lady.
lg = large (usually figures above 3 ft 6 in.).
mt = damaged.
pal = palimpsest.
Shr = figure in a shroud.
Skl = skeleton.
sm = small (usually figures under 1 ft 6 in.).
Sts = saints.
tr can = triple canopy.
W = wife or wives (e.g. 2W = two wives).
wn = worn down.
†† = an outstanding brass, usually very large.
† = a brass of fine quality, usually quite large.
* = an unusual, unique or curious brass, often quite small.

Children *with their parents* are not listed, except those in swaddling-clothes, who are denoted by the word 'Infant'. No attempt has been made to show whether palimpsests are mounted on hinges or fastened down, and only in particular cases has the reverse side been given.

BEDFORDSHIRE

Ampthill. C & W 1450; L sm wn 1485; C & W sm wn 1406; A 1532.
Aspley Guise. E in ac kng St. 1410†; A 1501.
Barford, Great. A & W 1525.
Barford, Little. C & W 1535.
Barton-in-the-Clay. E demi 1396; C sm wn 1490.
Bedford, St. Mary. C 1627; L 1663.
Bedford, St Paul. A & W 1573.
Biddenham. 2C & W 1490; 2Shr 1520; L 1639.
Biggleswade. C mt & W 1449; can Sts mt 1481.
Blunham. C & W 1506.
Bromham. A & 2W can lg pal 1435††.
Caddington. C & W 1505; C & 2W 1518.
Campton. C & W sm 1489.
Cardington. A & 2W her 1540; A & W 1638.
Clifton. A & W 1528.
Cople. Serjeant-at-law & W 1410*; A 1415; A & W 1435; A & W 1507; Judge & W her 1544; A & W kng 1556; Baron of Exchequer & W 1563.
Dean. E 1501.
Dunstable. C & W 1450; C & W 1502; C & W 1516; C & parents 1640.
Eaton-Bray. L 1501.
Eaton-Socon. C & W 1400; L sm wn 1450.
Elstow. L 1427; Abbess 1520†.
Eyworth. C & W 1624.
Felmersham. C & W sm wn 1610.
Flitton. L mt 1434; L 1544; A 1545; C 1628.
Goldington. C 1507; A 1585.
Gravenhurst, Lower. C & 3W 1606.
Hatley Cockayne. A 1430; L 1480; A & W 1515; A & 2W wn 1527.
Hawnes. C 1568.
Higham Gobion. C 1568.
Houghton Conquest. A, W & son A 1493; A & W 1500.
Houghton-Regis. E demi 1410; E 1506.
Husborne Crawley. C & W sm wn 1600.
Langford. E wn 1520.
Leighton Buzzard. 3C 1597; C & W 1636.
Lidlington, Old Church. C & W 1495.
Luton. C & son E 1415; L 1455; L tr can mt 1490; C 1500; A & 2W 1513; A & 2W 1513; L 1515; E 1515; C mt & 2W 1520; L 1593.
Marston Mortaine. E demi 1420; A & W 1451.
Maulden. A & W 1576; L kng sm 1594.
Mepshall. A wn 1440; A & W mt wn 1441.
Poddington. C 1518.
Pottesgrove. C & W 1535; C mt & W mt pal 1563.
Pulloxhill. A & W kng 1608.
Renhold. C & W sm 1518.
Salford. C & W 1505.
Sharnbrook. C & W 1522.
Shillington. E lg 1400†; E sm wn 1485.
Souldrop. C kng 1608.
Stagsden. A & W 1617.
Stevington. A 1422.
Sutton. Cross 1516.
Thurleigh. A wn 1420.
Tilbrook. C & W 1400.
Tingrith. C & 2 Infants 1611.
Toddington. Shr mt & Infant 1480.
Totternhoe. E 1524; C sm 1621.
Turbey. C sm 1480; E in ac 1500; L 1606.
Willshampstead. E demi 1450.
Woburn. Tr can mt 1394.
Wymington. C & W can 1391††; L 1407; A lg 1430†; E 1410.
Yelden. E 1433; E 1617; C 1628.

BERKSHIRE

Abingdon, St Helen. C demi 1417; E in ac wn 1501.
Appleton. Shr 1518.
Ashbury. C demi 1360; E mt 1409; E 1448.
Basildon. C & W sm 1497.
Binfield. E demi 1361.
Bisham. C 1517; C & 2W 1581.
Blewbury. E 1496; A & 2W 1515; A & W her 1523; A & W her 1548.

Bray. A & 2W her on bracket 1378†; Justice 1475; C & 2W 1490; C & W kng & infant 1610; C & W 1621.
Brightwalton. C sm 1517.
Brightwell. E 1507; C & W 1509; C & W sm 1512.
Buckland. C & W 1578.
Burghfield. A & W 1568.
Buscot. C mt & W 1500.
Childrey. A & W her can lg 1444†; E 1480; C & W 1480; E mt sm 1490; Shr & Trinity 1477; A & W & Trinity 1514; 2 Shr 1516; C & W & Trinity 1520; E in ac 1529.
Cholsey. E sm wn 1471.
Compton Parva. C & W 1520.
Cookham. C sm 1458; L & 2C sm 1503; A & W & Trinity sm 1510; A & W kng & Trinity 1527; C & W mt pal 1577.
Coxwell, Great. C 1509; L 1510.
Cumnor. A & W kng 1572; L mt 1577; C & W sm 1599.
Denchworth. A & W 1516; A & W pal 1562; A & W 1567.
Easthampstead. C demi 1443.
Faringdon. A mt & W 1443; C & W wn 1485; E sm 1505; A & 2W her kng mt 1547.
Fawley. L kng 1621.
Finchampstead. L 1635.
Hagbourne, East. A & W kng 1627.
Hanney, West. E mt lg 1370; A 1557; C & 2W 1592; A 1599; A & 2W 1602; C & W 1611.
Harwell. C & W 1599.
Hendred, East. C 1439; A & W 1589.
Hurst. C & W mt kng 1574; L in bed 1600*.
Kintbury. C & W 1626.
Lambourne. C & W demi 1406; 2C demi 1410; A her & Trinity 1485; C & W 1619.
Langford. C & W 1609.
Letcomb Regis. L mt sm 1440.
Lockinge, East. C & W 1624; L 1628.
Longworth. E demi 1422; 2Shr 1500; Shr mt 1509; L 1566.
Marcham. A & W kng 1540.
Mortimer, Stratfield. A 1441; L mt 1441.
Newbury. C & W & St 1519.
Reading, St Giles. C & W wn 1521.
Reading, St Laurence. C & W demi 1415; C pal 1538; C & W 1584.
Remenham. A mt sm 1591; E 1622.
Sandhurst. C & W sm 1608.
Shefford, Little. A & W kng 1524.
Shottesbrooke. E & C can lg 1370††; L 1401; A 1511; C & 3W 1567.
Sonning. A 1434; C sm 1546; C & W 1549; L 1575; L 1585; L sm 1627.
Sparsholt. E in cross mt 1353; C mt 1495; L sm 1510; C & W 1602.
Stanford-Dingley. L 1444; C 1610; C kng sm 1620.
Stanford-in-the-Vale. E demi lg 1398.
Steventon. C & W sm 1476; A & W 1584.
Streatley. L 1570; C sm 1583; C & W 1603.
Swallowfield. L sm 1466; A & W 1554.
Tidmarsh. L 1499; A her mt 1517.
Tilehurst. C & W sm 1469.
Ufton Nervet. C & W sm 1627.
Waltham, White. L mt 1445; L sm 1506.
Wantage. E demi 1370; A lg 1414†; E in ac 1512; C & 2W 1522; L 1619.
Warfield. C 1592.
Welford. E in ac sm 1490; C 1530.
Windsor, Old. C & 2W 1621.
Windsor, St George's Chapel. Can 1380; A & W her kng & Trinity 1475; E kng & Sts 1522†; Child in cradle 1630*; Child in cradle 1633*.
Winkfield. Yeoman of the Guard demi 1630*.
Wittenham, Little. E 1433; C sm 1454; L 1472; C1483; A kng 1588; Child 1683.
Wokingham. C mt & W 1520; C & W kng 1610.
Wytham. A & W mt 1455.

BUCKINGHAMSHIRE

Amersham. C & W 1430; C & W 1439; C mt wn 1450; C & W 1521; Child kng sm 1623.

Astwood. C & 2W 1534.
Beachampton. C 1600; L 1611.
Beaconsfield. C & W 1609.
Bledlow. E sm 1525.
Bletchley. E demi 1616.
Bradenham. E 1521.
Burnham. C & W sm 1500; C 1563; C & 3W 1581.
Calverton. C & W 1519.
Chalfont St Giles. E sm 1470; L 1510; C & 2W 1540; A & W 1565; A & 2W 1570.
Chalfont St Peter. A & W 1446; A & W 1446; E sm 1440.
Chearsley. C & W 1462.
Chenies. C & 2W wn 1469; A & W can wn 1484; E sm wn 1494; L can wn 1510†; L 1511; L 1524.
Chesham Bois. L 1516; A 1552; Infant 1520.
Chicheley. A mt & W 1558; Shr 1560.
Claydon, Middle. L 1523; E demi 1526; A & W 1542†.
Clifton Reynes. A mt 1428; 2 Shr 1500.
Crawley, North. E kng 1589.
Crendon, Long. C & W 1468.
Datchet. C & W 1593.
Denham. A & 2W wn 1494; Abbess 1540†; L pal 1545, on reverse Friar 1440*; E 1560.
Dinton. A & W 1424; C & W mt 1486; A & W 1551; a mt & W mt 1551; C & W 1558; A & W 1628.
Drayton Beauchamp. A lg 1368†; A lg 1375†; E mt sm 1531.
Dunton. C & W sm 1420; L sm 1510.
Edlesborough. E lg 1395†; inscribed Rose 1412*; A & 3W 1540; C & W 1592.
Ellesborough. A & W 1554.
Emberton. E 1410.
Eton College Chapel. E sm 1489; E tr can 1503†; E sm 1509; A sm 1521; E 1522; E in ac 1525; E 1525; L 1528; E with Garter 1540†; E demi 1545; L pal 1560; E in ac kng 1636.
Haddenham. E wn 1420; E demi 1428.
Halton. Baron of Exchequer & W kng 1553.
Hambleden. C demi 1457; C & W 1497; L kng 1500; C & W mt 1600; C 1634.
Hampden, Great. A & W 1525; A & 2W 1553.
Hanslope. Child 1602.
Hardmead. C mt 1556.
Haversham. L 1427; Skl 1605.
Hedgerley. C & W sm 1498; L pal 1540.
Hitcham. A & W 1510; A 1551.
Horwood, Great. E in ac sm wn 1487.
Hughenden. E sm 1493.
Iver. A & W 1508.
Ivinghoe. C & W sm 1517; C 1531; C 1576; C 1594.
Langley Marsh. C 1608.
Leckhampstead. L 1500; C 1506.
Lillingstone Dayrell. A & W 1491; E mt sm 1493.
Lillingstone Lovell. Inscribed Heart 1446*; C & W sm 1460; C & W 1513.
Linford, Great. C & W 1473; C & W 1536; C & W sm 1611.
Linslade. C & 3W 1500.
Loughton. E demi 1514.
Ludgershall. L 1600.
Marlow, Little. L 1430.
Marston, North. C 1602.
Marsworth. A mt pal 1586; L & Infant 1606; A, W & Infants with Death 1618*.
Milton Keynes. E 1427.
Missenden, Great. L sm 1510.
Missenden, Little. C 1613.
Moulsoe. A & W 1528.
Mursley. L 1570.
Nettleden. A 1545.
Newport Pagnell. C wn 1440.
Penn. Shr 1540; A & W mt 1597; A & W 1638; L 1640; A & W 1641.
Pitstone. L sm 1320*.
Quainton. L demi sm 1360; E in ac kng sm 1422; E 1485; L 1509; C 1510; L sm 1593.
Radnage. C 1534.
Risborough, Monk's. E 1431; C & W demi 1460.
Saunderton. L demi sm 1430.

Shalston. L 1540.
Slapton. E demi 1462; Yeoman of the Crown & 2W 1519; E sm 1529.
Soulbury. C & W 1502; L 1516.
Stokenchurch. A 1410; A 1415; C 1632; L 1632.
Stoke Poges. A & W 1425†; C & W 1577.
Stone. Shr mt & L 1472; C & W pal 1520.
Stowe. L sm 1479; Child 1592.
Swanbourn. C & W 1626.
Taplow. C in Cross 1350†; C, L & Shr 1455†; A & 2W 1540.
Thornborough. C & W 1420.
Thornton. A & 3W can 1472†; L 1557.
Tingewick. E kng 1608.
Turweston. E 1450; C & 2W sm 1490.
Twyford. E demi 1413; A pal 1550.
Tyringham. A 1484; L 1508.
Upton. Shr kng 1472; A & 3W 1517; A & W 1599.
Waddesdon. A tr can mt 1490; E sm 1543; Shr 1548, A & W 1561.
Wendover. C, W & 3Shr kng 1537.
Weston Turville. C 1580.
Weston Underwood. L mt 1571.
Whaddon. Serjeant-at-law & 2W 1519; L kng 1612.
Winchendon, Nether. A 1420; L 1420; C mt & W sm 1487.
Winchendon, Over. Canon 1502†.
Wing. C & W wn 1489; C & W wn 1490; C kng 1648.
Winslow. C & W sm 1578; L wn 1634.
Wooburn. C 1488; C & W 1500; E 1519; Shr & Trinity 1520; Child 1642.
Worminghall. C, W & Infant kng 1592.
Wotton Underwood. C, W & Infant 1587.
Wraysbury. A & W can 1488; Child sm 1512.

CAMBRIDGESHIRE
Abington-Piggotts. C 1460.
Balsham. E tr can Sts Trinity etc lg 1401††; E can Sts lg wn 1462†; A wn 1480.
Barton. C & W sm 1600.
Bassingbourn. C & W 1683.
Burwell. E tr can mt pal 1542.
Cambridge Colleges:
Christ's. A & W 1520; E in ac 1540.
Gonville & Caius. A wn 1500.
King's. E in ac 1496; E in ac 1507; E 1528; E 1558.
Queens'. E mt 1480; E in ac 1535; C 1591.
St John's. E in ac mt tr can lg wn 1414; E wn 1430.
Trinity Hall. E Sts lg 1510†; E in ac 1530; C 1598.
Cambridge Churches:
St Bene't. E in ac kng 1442.
St Mary-the-Less. E in ac mt 1436; E in ac demi 1500.
Croxton. C 1589.
Drayton, Dry. A & W 1540.
Ely Cathedral. Bishop lg 1554†; E in ac lg 1614†.
Fordham. C & W wn 1521.
Fulbourne. E 1390; E can lg 1391†; L 1470; E 1477; L 1480.
Girton. E 1492; E 1497.
Haddenham. C & W mt 1454.
Hatley, East. L 1520.
Hatley St George. A 1425.
Hildersham. C & W kng Trinity in Cross 1408††; A & W sm 1427; A can Trinity 1466†; Skl 1530.
Hinxton. A & 2W lg wn 1416.
Horseheath. A lg 1365†; C mt 1552.
Impington. A & W her 1505.
Isleham. A & W can mt 1451; A & 2W can mt 1484†; C & W lg 1574.
Kirtling. C kng 1553.
Linton. A 1424.
March. C & W wn 1501; A mt & W her kng Annunciation 1517*.
Milton. Justice & W 1553; C 1660.
Sawston. C mt 1420; A mt 1480; 2Shr 1500; E sm 1527.
Shelford, Great. E mt can mt 1418.
Shelford, Little. A & W 1410; A & W 1410; E in ac 1480.
Stapleford. E sm 1617.
Stow-cum-Quy. A 1460.

Stretham. L 1497.
Swaffham Prior. A & W sm 1462; C & W 1515; C & W 1521; C & W 1530; C 1638.
Trumpington. A lg 1289††.
Westley Waterless. A & W lg 1325††.
Weston Colville. L & Son mt 1427; E & W kng Angel, skulls etc. 1636.
Wicken. L sm 1414; C sm 1520.
Wilbraham. E in ac kng sm 1521.
Wilburton. E can lg 1477†; C & W 1506; C & W 1516.
Wimpole. E lg & B.V. Mary 1501; C 1500; L 1535.
Wisbech. A lg wn 1401†.
Wood Ditton. A & W mt 1393.

CHESHIRE
Chester, St Peter. Lawyer wn 1460.
Macclesfield. C & W kng, Mass of St Gregory 1506*.
Middlewich. L wn 1591.
Over. A 1510.
Wilmslow. A & W can mt wn 1460†.
Wybunbury. A & W 1513.

CORNWALL
Anthony, East. L can 1420†.
Blisland. E 1410.
Budock. A & W 1567.
Callington. Justice & W 1465.
Cardynham. E sm 1400.
Colan. C & W 1572; A & W 1575.
Constantine. C & W pal 1574; C & W kng 1616.
Crowan. A mt 1420; A & W 1490; A & W 1550; A mt 1599.
Fowey. C & W 1450; C 1450; C 1582; L 1602.
Goran. L kng 1510.
Grade. A & W 1522.
Helston. C & W 1606.
Illogan. A & W 1603.
Landrake. A sm 1509.
Lanteglos-near-Fowey. A 1440; A & W 1525.
Launceston. L 1620.
Lotswithiel. A wn 1423.
Madron. C & W 1623.
Mawgan-in-Pyder. E sm 1420; A & W 1573; L 1578; C 1580; Rose 1586*.
Minster. Child 1602.
Probus. C & W 1514.
Quethioc. C & W 1471; C & W 1631.
St Breock. C & W 1510.
St Columb Major. A & 2W 1545; A & W 1633; A & W 1633.
St Erme. C & W kng 1596.
St Glovias. C & W 1485.
St Ives. L kng, St Michael wn 1462.
St Just-in-Roseland. E 1520.
St Mellion. A & W 1551.
St Michael Penkivel. A 1497; E in ac 1515; C & W 1619; L 1622; A kng 1634.
St Minver. C 1517.
Sithney. Inscribed Cross mt 1420.
Stratton. A & 2W 1561.
Tintagel. L demi 1430.
Truro Cathedral. C 1585; C 1630.
Wendron. E mt 1535; C & W mt 1580.

CUMBERLAND
Arthuret. Cross & Heart 15th cent.
Bootle. A 1562.
Carlisle Cathedral. Bishop tr can lg 1496†; Bishop kng 1616*.
Crossthwaite. A & W 1527.
Edenhall. A & W her 1458.
Greystoke. E demi sm 1526; L sm 1547; C sm 1551.

DERBYSHIRE
Ashbourne. A & W her tr can 1538†.
Ashover. E 1504; A & W 1507.
Bakewell. C sm 1648.
Beeley. Shr 1710.
Chesterfield. A & W her 1529.
Crich. Infant 1637.
Dronfield. 2E 1399; C & W 1580.
Edensor. A 1570.
Etwall. L Sts 1512; A & 2W her kng 1557.
Hathersage. A & W 1463; C & W 1493; A & W her 1500; A & W her kng 1560.
Hope. C 1635.
Kedleston. A & W mt 1496.

Longstone, Great. C & W kng 1624.
Morley. A & W kng, St Christopher 1454; A & 2W Sts 1470†; A & 3W 1481; A & W kng, St Christopher 1525; A & W 1558.
Mugginton. A & W 1475†.
Norbury. Justice mt & W her pal 1538.
Sawley. A & W 1467; A & W 1478; C mt & W 1510.
Staveley. A mt Trinity 1480; A & W kng B.V. Mary 1503.
Teddington. C & W 1505.
Tideswell. Trinity 1462; C & W 1500; Bishop 1579†.
Walton-on-Trent. E 1492.
Wilne. A, W & Son her kng Trinity 1513.
Wirksworth. C & W 1510; C & W 1525.
Youlgreave. L 1604.

DEVONSHIRE
Allington, East. L kng mt 1540; C & W 1595.
Atherington. A & 2W 1539.
Bigbury. L 1440; L 1460.
Blackawton. C & W 1582.
Braunton. L kng pal 1548.
Chittlehampton. C & 2W 1480.
Clovelly. A 1540; A 1540.
Clyst St George. L kng 1614.
Dartmouth, St Petrock. C lg 1609; L 1610; L sm 1617.
Dartmouth, St Saviour. A & 2W can lg 1408†; L sm 1470; C 1637.
Ermington. C & W kng 1583.
Exeter Cathedral. A can wn 1409; E kng 1413.
Filleigh. A kng 1570; A kng 1570.
Haccombe. A 1469; A 1586; L 1589; L 1611; A & W kng 1656.
Harford. A 1566; C & W 1639.
Hartland. L kng 1610.
Monkleigh. 2 Angels holding scroll 1509*; A kng 1566.
Ottery St Mary. 3C 1620.
Petrockstow. A & W kng 1591.
St Giles-in-the-Wood, nr. Torrington. L mt 1430; L 1592; L kng 1610.
Stampford Peverell. L 1602.
Sandford, nr. Crediton. L 1604.
Shillingford. A & W her 1499.
Staverton. C demi 1592.
Stoke-Fleming. C & Granddaughter can mt lg 1391†.
Stoke-in-Teignhead. E 1370.
Tedburn St Mary. E in ac kng 1580; E & W 1613.
Tiverton. C & W lg 1529.
Tor Mohun. L 1581.
Ugborough. L 1500.
Washfield. C & W 1606.
Yealmpton. A 1508.

DORSETSHIRE
Bere Regis. C & W kng 1596.
Caundlepurse. A 1500; L sm 1527; E mt sm 1536.
Compton Valence. E demi wn 1440.
Corfe Mullen. C wn 1437.
Crichel More. L 1572.
Evershot. E 1424.
Fleet, Old Church. A & W kng 1603; A & W kng 1612.
Knowle. A & 2W kng 1572.
Langton, nr. Blandford. C & 2W 1467.
Lychett Matravers. Shr sm 1470.
Milton Abbey. A her kng 1565.
Moreton. A kng 1523.
Piddlehinton. E 1617.
Piddletown. C mt 1517; A her kng Trinity 1524; A & W kng 1593.
Pimperne. Skl 1694.
Puncknowle. A kng 1600.
Rampisham. C & W 1523.
Shapwick. L sm 1440; E sm 1520.
Sturminster-Marshall. E sm 1581.
Swanage. 2L sm 1490.
Thorncombe. C & W lg 1437†.
Wimborne Minster. St Ethelred demi with Crown, engraved 1440*.
Woolland. L kng 1616.
Yetminster. A & W 1539.

DURHAM
Auckland St Andrew. E mt 1380; L sm 1581.

Auckland St Helen. C mt & W mt 1470.
Billingham. E mt 1485.
Brancepeth. A wn 1400; E in ac demi wn 1456.
Chester-le-Street. L 1430.
Hartlepool. L 1593.
Haughton-le-Skerne. L holding twin Infants 1592*.
Houghton-le-Spring. L kng 1587.
Sedgefield. 2 Shr 1500.

ESSEX
Althorne. B.V. Mary 1502; C Trinity 1508.
Arkesden. A 1439.
Ashen. C & W 1440.
Aveley. A can sm fgn 1370*; Infant mt 1583; 2 Children 1588.
Baddow, Great. L 1614.
Bardfield, Great. L 1584.
Barking. E in ac wn 1480; E 1485; C & W 1493; C, W & Infants 1596.
Belchamp St Paul. A 1587.
Bentley, Little. A mt & W 1490.
Berden. C & 2W 1473; C & W 1607.
Blackmore. C mt 1420.
Bocking. A & W 1420; C 1613.
Boreham. L 1573.
Bowers Gifford. A mt lg 1348.
Bradfield. L 1598.
Bradwell-on-the-Sea. L 1526.
Braxted. A & 2W 1508.
Brightlingsea. C & W 1496; L 1505; L 1514; C & 2W 1521; C & W 1525; 2L 1536 on bracket 1420; C 1578.
Bromley, Great. E can 1432.
Canfield, Great. L 1530; A & W kng 1558; A & W 1558.
Canfield, Little. 2L 1578; L 1593.
Chesterford, Great. L 1530; Infant 1600.
Chesterford, Little. L 1462.
Chigwell. Archbishop lg 1631†.
Chrishall. A & W tr can 1380††; L sm 1450; C & W kng wn 1480.
Clavering. C mt & W 1480; C & W 1591; C & W 1593.
Coggeshall. 2L 1490; C & W 1520; C & W 1533; C 1580.
Colchester, St James. C 1569; L mt pal 1584.
Colchester, St Peter. Alderman & W kng 1530; L, C & Alderman kng 1533; C kng 1563; C & W kng 1572; C & 2W kng 1610.
Cold Norton. L 1520.
Corringham. E demi 1340; C mt wn 1460.
Cressing. L & Infant 1610.
Dagenham. Baron of Exchequer & W 1479.
Dengie. L mt 1520.
Donyland East, New Church. C 1621; L 1627.
Dovercourt. C 1430.
Dunmow, Great. L 1579.
Easter, Good. L 1610.
Easton, Little. E 1420; A in robes of Garter & W 1483††.
Eastwood. C 1600.
Elmdon. C & 2W 1530.
Elmstead. Heart 1500.
Elsenham. L kng 1615; L kng 1619.
Epping, Old Church. E in ac 1621.
Fambridge, North. C & W mt 1590.
Faulkbourne. A 1576; L 1598.
Felsted. A 1415; L demi 1420.
Finchingfield. A & W her 1523.
Fingringhoe. C & W 1600.
Fryerning. L pal 1563.
Goldhanger. L 1531.
Gosfield. Serjeant-at-Law 1440†.
Halstead. A & 2W lg 1420†; L kng & Infant 1604.
Ham, East. L 1610; L 1622.
Ham, West. C & 4W 1592.
Hanningfield, West. L demi 1361.
Harlow. A & W sm 1430; C & W 1490; C & W 1518; C mt sm 1569; C & W 1582; C & W mt 1585; C & Death sm 1602; C sm 1615; A & 2W 1636; C & W kng 1642.
Hatfield Broadoak. Head of L 1395.
Hatfield Peverell. C & W kng 1572.
Hempstead. C & W 1475; C 1480; A & W 1498; C & W 1518; C & W sm 1530.

Henny, Great. C & W 1530.
Heybridge. C 1627.
Horkesley, Little. A tr can & Son A tr can lg 1412††; Shr 1502; L & 2A her 1549.
Hornchurch. C & W sm 1591; 2L 1602; C & W sm 1604.
Horndon, East. L 1476; A mt 1520.
Hutton. A & W 1525.
Ilford, Little. Schoolboy 1517*; Infant 1630.
Ingrave. L 1466; A & 4W her 1528.
Kelvedon Hatch, Old Church. L sm 1560.
Laindon. E 1470; E sm 1510.
Lambourne. C & W 1546.
Latton. Baron of Exchequer & W 1467; A & W 1490; C & W 1600; L 1604.
Laver, High. A & W 1496.
Leigh, nr. Rochford. 2C & 2W 1453; C & W 1632; C & W 1640.
Leighs, Great. Head of E 1370; E demi mt 1414.
Leyton, Low. L sm 1493; C & W 1620.
Lindsell. C & W 1514.
Littlebury. C 1480; E 1510; C & W 1510; C 1520; L 1578; L 1624.
Loughton. C & 2W 1541; A & W can 1558; C 1594; C & W kng 1637.
Margaretting. A mt & W 1550.
Matching. C & W 1638.
Messing. L 1540.
Netteswell. C & W 1522; C, W & Infant 1607.
Newport. C & W 1515; C & W 1608.
Noak Hill, nr. Romford. C 1450; C 1480: C 1600. (Brasses moved to South Weald.)
Ockendon, North. A & W 1502; L 1532.
Ockendon, South. A mt can mt 1400; L 1602.
Ongar, High. C 1510.
Orsett. C kng sm 1535.
Parndon, Great. C 1598.
Pebmarsh. A mt lg 1323†.
Rainham. L 1480; C & W 1500.
Rawreth. A & W kng 1576.
Rayleigh. C mt & W wn 1450.
Rettendon. C & 2W 1535; C 1605; C 1607.
Rochford. L sm 1514.
Roydon. A & W 1471; A & 2W her 1521; C 1570; L 1589.
Runwell. A & W kng 1587.
Saffron Walden. E, Pelican in Piety 1430; 2L 1480; L 1490; L 1500; C 1510; L 1530; C mt 1530; C 1530.
Saffron Walden Almshouses. 3C & 2L 1475.
Sandon. E in ac & W kng 1588.
Shopland. Serjeant-at-Arms mt 1371. (Brass moved to Sutton.)
Southminster. C & W 1560; C 1634.
Springfield. A 1421.
Stanford Rivers. Infant 1492; A & W 1503; A & W 1540; L kng 1584.
Stebbing. L 1390.
Stifford. E demi 1378; Shr 1480; C & W 1504; C & W 1622; L sm 1627; L sm 1630.
Stisted. L kng 1584.
Stock. A 1547.
Stondon Massey. C & 2W 1570; A & W pal 1573.
Stow Maries. L 1602.
Strethall. E in ac 1480.
Terling. A & W 1500; C & W kng 1584; C & 2w kng 1584.
Thaxted. E in ac 1450.
Theydon Gernon. E 1458; A & W kng 1520; L kng 1567.
Thorrington. L 1564.
Thurrock, Grays. C & 2W 1510.
Thurrock, West. 2C 1585.
Tillingham. C kng 1584.
Tilty. A & W lg 1520; A & W 1562; L kng Infants sm 1590.
Tollesbury. C & W 1517.
Tolleshunt Darcy. Border fgn Sts pal 14th cent; A mt & W 1420; L pal 1535; A 1540; L 1559.
Toppesfield. C & W 1534.
Totham, Great. L 1606.
Twinstead. C & W 1610.
Upminster. L 1455; C pal 1530; C & W pal 1545; L 1560; A 1591, L 1626.

Waltham Abbey. C & W kng 1565; C & W kng 1576.
Waltham, Great. C & W 1580; C 1580; C & W 1617.
Waltham, Little. A 1447.
Walthamstow. C & W kng 1543; C & W mt pal 1588.
Warley, Little. L demi 1592.
Weald, North. C & W 1606.
Weald, South. Justice mt kng 1567; 2 Children kng 1634.
Wenden Lofts. C & W 1460.
Wendens Ambo. A 1410.
Widdington. C 1450.
Willingale-Doe. A 1442; L 1582; L 1613.
Wimbish. A mt & W sm in Cross mt 1347†.
Wivenhoe. A tr can lg 1507†; E sm 1535; L her tr can lg 1537†.
Woodham Mortimer. Child mt 1584.
Woodham Walter. Death's Head 1650.
Wormingford. C 1460; C & 2W 1590.
Writtle. A & W 1500; C & 4W 1510; L 1513; L sm 1524; C & W 1576; W 1592; C & W kng 1606; C 1609; W 1616.
Yeldham, Great. C & W kng 1612.

GLOUCESTERSHIRE

Abbenhall. C & W 1609.
Berkeley. C mt 1526.
Bibury. Skl 1707; Skl 1717.
Bisley. L 1515.
Bristol Grammar School. C & 2W kng 1570.
Bristol Churches:
St James. C & W kng 1636.
St John. C & W 1478.
St Mary Redcliffe. Baron of Exchequer 1439; A & 2W her kng 1475; C & W can 1480†; Serjeant-at-law & W 1522.
St Stephen. C & W kng 1594.
St Werburgh, New Church. C & W kng 1586.
Trinity Almshouse Chapel. C sm can 1411; L sm can 1411.
Cheltenham, St Mary. Justice & W wn 1513.
Chipping Camden. C & W can lg 1401††; C & W 1450; C & W 1467; C & 3W 1484.
Cirencester. C mt & W mt on wine casks can mt lg wn 1400; A can lg 1438; C on woolsack & W can 1440†; C & 4W 1442; Angel & Lily-pot wn 15th cent; A & 2W 1462; C mt & W 1470; E wn 1478; C mt & W wn 1480; L mt 1480; C mt & 2W 1497; C 1500; C & W kng 1500; 2L 1530; C & Shears 1587; C & W 1626.
Clifford Chambers. A & W 1583; L & Infant 1601.
Coaley. E & W kng 1630.
Deerhurst. Baron of Exchequer & W can St lg 1400††; L mt 1520; L 1525.
Dowdeswell. E 1520.
Doynton. C & W 1529.
Dyrham. A & W can mt lg 1401†.
Eastington. L her 1518.
Fairford. A & W 1500; A & 2W her 1534; A & 2W kng Trinity 1534.
Gloucester:
St John Baptist. C & W mt 1520.
St Mary de Crypt. Alderman & W tr can St John Bapt. 1529†.
St Michael. 2L 1519.
Kempsford. C & W 1521.
Lechlade. C on woolsack & W 1450; C 1510.
Leckhampton. C & W kng 1598.
Micheldean. 2L 1500.
Minchinhampton. C & W 1500; 2Shr 1510; C & W 1519.
Newent. A sm 1523.
Newland. A mt & W, miner with candle, bag & pick 1443*.
Northleach. C on woolsack & W lg 1400†; L mt, C mt & C on shears can wn 1447*; C on sheep & woolsack can mt lg 1458††; C & W 1485; C & W, sheep on woolsack 1490; C & W 1501; C on woolsack & W can 1526*; E kng 1530.
Olveston. 2A her kng 1506.

Quinton. L can 1430.
Rodmarton. Lawyer 1461.
Sevenhampton. C 1497.
Thornbury. L 1571.
Toddenham. C & W 1614.
Tormarton. C 1493.
Weston-sub-Edge. C 1590.
Weston-upon-Avon. A her 1546; A her 1559.
Whittington. C, W & Infant 1560.
Winterbourne. L 1370.
Wormington. L in bed & Infant 1605★.
Wotton-under-Edge. A & W lg 1392†.
Yate. C & 2W 1590.

HAMPSHIRE
Alton. L 1510.
Basingstoke. C & W 1606; Child 1621.
Bramley. C wn 1452; L 1504; C & W 1529.
Bramshott. C & W 1430.
Candover, Brown. C, L & A 1490.
Candover, Preston, Old Church. L 1607.
Crondall. E lg 1381†; A kng 1563; Skl 1641.
Dean, Prior's. C & W 1605.
Dogmersfield. L kng & Infant 1590.
Dummer. C kng 1580.
Eversley. Cross 1502★.
Fordingbridge. A & W kng 1568.
Froyle. C 1575.
Havant. E 1413†.
Headbourne Worthy. C sm 1434.
Headley. C & W sm 1510.
Heckfield. L 1514.
Kimpton. A & 2W kng Cross 1522.
Kingsclere. L sm wn 1503; E sm 1519.
Mapledurwell. C & W sm wn 1520.
Monxton. L & C kng 1599.
Oakley, Church. C & W 1487.
Odiham. C & W 1480; E 1498; L 1504; L 1522; C 1530; A mt & 2W 1540; Infant 1636.
Ringwood. E Sts can mt lg 1416†.
Sherborne St John. C & L demi sm 1360; A her kng 1488; Shr 1488; A & 2W 1492; A kng Trinity 1492; A kng sm 1540.
Sherfield-on-Loddon. L kng 1595; C kng 1600.
Sombourne, King's. 2C 1380.
Southampton, God's House. E mt 1500.
Southwick. A & W 1548.
Stoke Charity. A & Resurrection 1482; A & W mt Trinity 1483.
Sutton, Bishop's. A mt & W 1520.
Thruxton. A tr can lg 1425††.
Tytherley, West. L 1480.
Wallop, Nether. Prioress 1436★.
Warnborough, South. A kng 1512.
Weeke. St Christopher sm 1494★.
Whitchurch. C & W 1603.
Winchester, College Chapel & Cloisters (Brasses in the Chapel disappeared in 1875 and are facsimiles). E 1413; E demi 1432; E demi 1445; E lg 1450; E demi 1473; E demi 1494; C 1498; E demi 1509; E demi 1514; E demi 1514; E in ac kng sm 1524; E 1548.
Winchester, St Cross. E lg 1382†; E 1493; E 1518.
Winchester, St Swithin. Infants 1612.
Yateley. C & W 1517; C & W 1532; L mt 1578; C 1590.

ISLE OF WIGHT
Arreton. A mt 1430.
Calbourne. A 1380.
Freshwater. A 1365.
Kingston. C 1535.
Shorwell. E 1518; 2L 1619.

HEREFORDSHIRE
Brampton Abbots. L sm 1506.
Burghill. C & W kng 1616.
Clehonger. A & W 1470.
Colwall. A & W 1590.
Hereford Cathedral. St Ethelbert holding his crowned head 1290★; Bishop can lg 1360††; E sm in Cross mt 1386†; C mt 1394; E wn 1434; A & W can lg 1435†; E can Sts 1476; C on cask wn 1480; A mt 1480; E mt 1490; A & 2W 1514; E sm 1520; Annunciation, Execution of John Bapt., Saints & Angels 1524★; E tr can Sts lg wn 1529; C & W 1600.

Kinnersley. E demi wn 1421.
Ledbury. E in ac kng 1410; A 1490; A 1614.
Llandinabo. Child in pond 1629*.
Ludford. A & W 1554.
Lugwardine. L kng 1622.
Marden. L & Infant 1614*.

HERTFORDSHIRE
Albury. A & W 1475; C & W 1588; A & W 1592.
Aldbury. C sm 1478; A & W her 1547.
Aldenham. C & W 1520; C mt 1520; C & W 1520; C 1520; C & 2W 1525; 2L mt 1525; L sm 1535; L sm 1538; Shr 1547; C & W 1608.
Amwell, Great. C mt & W wn 1490.
Ardeley. L mt 1420; E 1515; C & W 1599.
Aspenden. C & W 1500; A & W her kng 1508.
Aston. Yeoman of Guard & W 1592.
Baldock. L mt 1410; C mt with horn, knife & rope & W sm 1420*; 2Shr 1480; C & W 1480.
Barkway. C & 2W 1561.
Barley. E 1621.
Bayford. C & W mt pal 1545; A 1630.
Bennington. E with badge mt 1420.
Berkhamstead, Great. C & W can wn 1356†; L 1370; E demi 1400; C 1485; Shr 1520.
Braughing. C & W pal 1480; C mt 1490; L 1561.
Broxbourne. E 1470; A mt & W her 1473†; E in ac 1510; Serjeant-at-arms & W 1531.
Buckland. L 1451; E 1478; C 1499.
Buntingford. E in pulpit 1620*.
Cheshunt. Tr can mt 1448; C mt & W wn 1449; L wn 1453; L 1502; L kng sm 1609.
Clothall. E 1404; E Trinity 1519; E 1535; L 1578; E 1602.
Digswell. A & W lg 1415†; A 1442; 2Shr 1484; C & W 1495; C & W 1530.
Eastwick. L 1564.
Essendon. C & W kng 1588.
Flamstead. E 1414; C & W 1470
Gaddesden, Great. C & W 1506; L 1525.
Hadham, Great. E in ac demi 1420; C & W 1520; C & W 1582; C & 2W 1610.
Hadham, Little. E sm wn 1470; A & W 1485.
Harpenden. C & W wn 1456; C & W 1571.
Hemel Hemstead. A & W 1390.
Hertford, All Saints. C mt 1435.
Hinxworth. C & W 1450; Alderman & W 1487.
Hitchin. C 1420; C mt & W wn 1421; C & W 1452; L wn 1470; Shr 1477; C & W 1481; Shr mt 1485; 2Shr 1490; 2Shr 1490; E 1498; C & W 1530; C & W 1535.
Holwell. Chalice, Wafer and Wild men 1515*.
Hunsdon. Shr & Trinity 1495; Park-keeper, Death & Stag 1591*.
Ickleford. C & W demi wn 1400.
Ippolyts. C & W kng 1594.
Kelshall. C & W wn 1435.
Kimpton. L 1450.
Knebworth. E Sts 1414†; A & 2W 1582.
Langley, Abbot's. 2L 1498; C & 2W 1607.
Langley, King's. L sm 1528; L pal 1578; C & 2W 1588.
Letchworth. C & W demi wn 1400; E 1475.
Mimms, North. E can Sts fgn 1380†; L 1458; A 1488; C & W 1490; A & W 1560.
Newnham. C & 2W 1490; L 1607.
Offley. C & 2W 1529; C & 3W 1530.
Pelham, Brent. 2L sm 1627.
Pelham, Furneux. C & W 1420†; A & W kng 1518.
Pelham, Stocking. Heart 1440.
Radwell. C & W 1487; C & 2W 1516; L 1602.
Redbourn. C mt & Peacock 1512; A & W kng 1560.
Rickmansworth. C & 2W 1613.
Royston. E in ac mt can mt 1421; Cross 1500; C & W 1500.

St Alban's Abbey. Abbot can Sts fgn lg 1370-80††; Abbot mt 1400; C & W wn 1411; Monk 1450; Tr can mt 1451; Monk 1460; A mt & W 1468; C 1468; C 1465; C mt 1470; Monk demi 1470; A 1480; C 1519; Sub-prior 1521.
St Alban's, St Michael. C & W 1380†; A 1380; C in Cross mt 1400.
St Alban's, St Peter. C & W 1627.
St Alban's, St Stephen. A & W 1482.
Sandon. A & W 1480.
Sawbridgeworth. A & W 1437; C & 2W 1470; 2Shr 1484; L 1527; A & W 1600; L 1600.
Shenley. C & W 1621.
Standon. C 1460; Alderman & A her 1477; A 1557.
Stansted Abbots. A 1490; C & W 1540; C 1581.
Stevenage. E mt 1500.
Tewin. C 1610.
Walkern. C & W 1480; C & W pal 1583; C & W wn 1636.
Ware. L wn 1400; L 1454; C & 2W 1470.
Watford. Justice mt lg 1415†; L lg 1416; 3C 1613.
Watton-at-Stone. A can lg 1361†; E lg 1370†; C 1450; L 1455; C mt 1470; A 1514; L mt 1545.
Wheathamstead. C & W 1450; A mt & W mt 1480; C & W 1520; C 1510; L 1510.
Willian. E 1446.
Wormley. C & W 1479; C mt & W Trinity, dogs hunting hare etc. 1490; C & W 1598.
Wyddiall. C mt 1532; C & W 1546; L demi 1575.

HUNTINGDONSHIRE
Broughton. C mt 1490.
Diddington. A mt & W her kng can Sts 1505; L 1513.
Godmanchester. C 1520.
Offord Darcy. A mt & 2W mt pal 1440; E in ac kng 1530.
Sawtry, All Saints. A & W lg 1404†.
Somersham. E 1430.
Stilton. C & W 1606; 2C sm 1618.
Stukeley, Little. C 1590.

KENT
Acrise. L 1601.
Addington. A mt 1378; A & W can 1409†; A 1415; A sm 1445; E demi 1446; A & W 1470.
Aldington. A & W 1475.
Appledore. L 1520.
Ash-next-Sandwich. L mt can mt 1455; L 1445; C & W wn 1525; A & W 1602; C & W 1642.
Ash-next-Wrotham. E demi 1465; E 1605.
Ashford. Head of E 1320; L mt can 1375†; Head of A & Angel 1499.
Aylesford. A & W 1436†.
Barham. C mt lg 1370; A & W 1455.
Bearsted. C & W kng 1634.
Beckenham. A & 2W her kng 1552; L 1563.
Bethersden. C pal 1459; C sm 1591.
Bexley. Hunting horn 15th cent; C sm 1513.
Biddenden. L & 2C 1520; A & W kng 1566; C mt pal 1572; C & 2W 1584; C & 3W 1593; C & 2W 1598; C & 2W 1609; C & W 1628; C & W 1641.
Birchington. 2C 1449; C 1454; L 1518; E 1532; L 1528; L & Infant mt 1533.
Birling. C 1522.
Bobbing. A & W can mt 1420; A mt tr can mt 1420; L sm 1496.
Borden. C 1450; E sm 1521.
Boughton Malherbe. C & W kng 1499; A & W 1529.
Boughton-under Blean. C & W sm 1508; C & W 1591; A 1587.
Boxley. E in ac 1451; A 1576.
Brabourne. A can mt lg 1433; L 1450; A 1524; L 1528.
Bredgar. E in ac 1518.
Brenchley. C & 3W 1517; C & W 1540.
Bromley. C & 2W 1600.
Brookland. E 1503.
Canterbury Churches
St. Alphege. E in ac 1523.
St. George. E 1438 (now in Cathedral Chapter Office).

St. Gregory. C kng 1522.
St. Martin. C & W 1587; A 1591.
Capel-le-Ferne. C & W sm 1526.
Challock. C & W sm 1504.
Chart, Great. Notary 1470; C & W 1495; C & 5W sm wn 1499; C & W sm wn 1500; A & 2W 1513; A & W 1565; A kng 1680.
Chartham. A mt lg 1306††; E sm 1416; E 1454; E 1508; L sm 1530.
Chelsfield. Crucifix mt & St Mary 1417; E sm 1420; E sm 1420; L 1480; L 1510.
Cheriton. E in ac sm 1474; E sm 1502; L 1592.
Chevening. E & W 1596.
Chistlehurst. E demi 1482.
Cliffe-at-Hoo. C & W 1609; C & 2W 1652.
Cobham. L can lg 1320††; A can lg 1354††; A holding Church can lg 1365††; A can lg 1367†; L can lg 1375†; L can lg 1380†; L can B. V. Mary lg 1395††; E on bracket tr can 1402†; A demi 1402; A can Trinity lg 1405††; A tr can Sts lg 1407††; E demi 1418; L lg 1433†; E in Cross mt 1447; E 1450; E 1498; L can Trinity 1506; A & W 1529.
Cowling. L sm 1508.
Cranbrook. C & Infant 1520; C & W kng 1627.
Cray, St Mary. C & 3W sm 1508; C & W 1604; L 1773*; C 1773*.
Cudham. L 1503.
Dartford. C & W can lg 1402†; L 1454; L 1464; C & W 1496; C & W 1508; L mt 1590; C & 2W, Infant 1590; L mt 1612.
Davington. C, W & Infant kng 1613; L kng 1616.
Deal, Upper. C & W 1508; A kng 1562; Infant 1606.
Ditton. A mt 1576.
Dover, St James. E & W 1613.
Dover, St Mary. C & W 1638.
Downe. C & W 1400; C sm 1420; C & W lg 1607.
Eastry. A & W 1590.
Edenbridge. C & W 1558.
Elmsted. L 1507.
Erith. C sm 1425; C & W 1435; L 1471; C & W 1511; A & W 1537.
Farningham. E demi 1451; L sm 1514; C kng sm 1517; C & W 1519.
Faversham. C mt can mt Trinity lg 1414; A mt 1419; E can mt 1480; L mt 1492; C mt wn 1496; C mt 1500; L mt wn 1500; C mt 1500; C 1510; E 1531; C & W can lg 1533; C & 2W wn 1533; C 1580; C mt 1610.
Fordwich. L 1605.
Goodnestone-next-Wingham. C & W Trinity 1507; L 1523; C & W 1558; A & W 1558.
Goudhurst. A can 1424; A 1490; A 1520.
Grain, Isle of. C 1520.
Graveney. C & L demi can mt 1360; A 1381; Justice & W can lg 1436††.
Halling, Lower. L in bed 1587.
Halstead. A 1444; C mt & W mt 1528.
Halstow, High E demi 1398; E mt & W mt sm 1618.
Hardres, Upper. E in ac kng Sts on bracket 1405††; A mt 1560.
Harrietsham. L kng Infant 1603.
Harty, Isle of Sheppey. C mt sm 1512.
Hawkhurst. C & W 1499.
Hayes. E demi 1460; E sm 1479; E sm 1523.
Headcorn. Child kng 1636.
Herne. A & W 1430†; E in ac 1450; L 1470; L 1539; C & 2W 1604.
Hever. L Angels 1419†; Cross sm 1520; A in Garter Robes lg 1538††; C kng 1585.
Hoath. L mt 1430; C & W sm wn 1532.
Hoo, All Hallows. A kng 1594.
Hoo, St Werburgh. E demi 1406; E mt lg 1412; C 1430; 2C 1436; A & W 1465; L 1615; C & W 1640.
Horsmonden. E can wn lg 1340†; L 1604.
Horton Kirby. L lg 1460; C & W 1595.
Hunton. C 1516.
Ightham. A & W her 1528; Ascension mt wn 1573; L 1626.
Iwade. C mt & W 1467.

Kemsing. E demi 1347.
Kingsnorth. A & W 1579.
Lee, St Margaret. L sm 1513; L pal 1582; A kng 1593.
Leeds. C & W 1509; L sm 1514.
Leigh. Shr L kng & Angel 1580*; C 1591.
Luddesdown. A mt 1450.
Lullingstone. A 1487 L sm 1533; L 1544.
Lydd. E in ac 1420; C 1429; C & W can 1430; C sm 1508; C 1520; C & W 1557; C & W 1566; C 1578; C 1590; L 1590; C 1608.
Lynsted. L 1567; C & W 1621.
Maidstone, All Saints. C, 2W & ancestors kng 1593*; E in ac & W kng 1640.
Malling, East. C & W 1479; E 1522.
Malling, West. C sm 1479; C sm 1532; L mt 1543.
Margate. C 1431*; Heart 1433; C & W 1441; C sm 1442; A sm 1445; Skl 1446; L mt 1475; E sm 1515; A 1590; Ship 1615*.
Mereworth. A mt can mt lg 1366; C & W 1479; C kng Christ 16th cent.
Mersham. E sm 1420.
Milton-next-Sittingbourne. A 1470; A & W her 1500; L kng 1529.
Minster, Isle of Sheppey. A lg 1330†; L 1335†.
Monkton, Isle of Thanet. E 1460.
Murston. A & W 1488.
Newington-next-Hythe. L mt 1480; Shr & W 1501; E sm 1501; C & 3W 1522; C 1570; A & W 1630.
Newington-next-Sittingbourne. 2C 1488; L 1580; C & 2W 1581; L 1600.
Northfleet. E lg 1375†; E demi 1391; A mt & W 1433.
Orpington. E 1511.
Otham. C & 3 W kng 1590.
Otterden. A 1408; L 1488; A 1502; A 1508; L 1606.
Peckham, East. C & W sm 1525.
Peckham, West. L 1465.
Pembury. Child 1607.
Penshurst. C & W 1514; Cross sm 1520.
Pluckley. A 17th cent; A wn 1440; A 17th cent; A 17th cent & W 1517; L 1526; A 17th cent; A her kng 17th cent; A & W kng 1610 (17th cent. brasses are restorations).
Preston-next-Faversham. A & W 1442; A 1459; L 1612.
Rainham. A sm 1514; C 1529; L 1530; C & W 1585.
Ringwould. C sm 1530.
Rochester, St Margaret. E demi pal 1465.
Romney, New. C 1510; C & W 1610.
Romney, Old. C & W 1526.
St Laurence, Thanet. A 1444; L 1493.
St Mary-in-the-Marsh. L 1499; C 1502.
St Nicholas-at-Wade, Thanet. C & 2W 1574.
St Peter, Thanet. C mt & W 1485; C & W sm wn 1503.
Saltwood. E demi 1370; A & W 1437; Angel and Heart 1496.
Sandwich, St Clement. C can mt wn 1490.
Seal. A lg 1395†; C pal 1577.
Selling. C 1525.
Sheldwich. A & W mt can 1394†; A & W 1426; Shr demi 1431.
Shorne. C & W demi 1457; L mt 1470; Chalice 1519*; L 1583.
Snodland. C mt sm 1441; C sm 1486; C & W sm 1487.
Southfleet. L on bracket 1414; C & W 1420; Cross mt 1420; E demi 1456; Shr sm 1520; C & W 1520.
Staple. C 1510.
Staplehurst. L 1580.
Stockbury. C & W 1617; L 1648.
Stoke. E mt 1415.
Stone. E sm in Cross 1408†.
Stourmouth. E in ac 1472.
Sundridge. A 1429; C 1460; A & W 1518.
Sutton, East. A & W lg 1629†.
Teynham. A 1444; C & Infant 1509; C sm 1533; C & W 1639.
Thanington. A 1485.
Tilmanstone. C & W kng 1598.
Trottescliffe. C & W 1483.
Tudeley. C & W 1457.
Tunstall. E 1525; L 1500.
Ulcombe. A can mt 1419; A 1442; A & W 1470†.

Upchurch. C & W demi 1350.
Westerham. C & 2W 1511; C 1529; C 1531; C 1533; C & 2W 1557; C & 2W 1566; E wn 1567.
Wickham, East. C & W demi sm in Cross 1325†; Yeoman of the Guard & 2W 1568.
Wickham, West. E sm 1407; E sm 1515.
Wittersham. C 1527.
Woodchurch. E sm in Cross 1330†; A & 2W kng 1558.
Wouldham. C & Child kng 1602.
Wrotham. C & W 1498; C 1500; A & W 1512; A & W her 1525; C 1532; A & W 1611; L 1615.
Wye. L & 2C 1440.

LANCASHIRE
Childwall. A & W her 1524.
Eccleston. E 1510.
Flixton. A & 2W Infants kng 1602.
Lancaster, St Mary. C 1639.
Manchester Cathedral. E can 1458; A mt & W mt 1460; Bishop mt 1515; A & W wn pal 1540; C & W kng 1607; C & W kng 1630.
Middleton. A & W 1510; E 1522; L & 2A 1531; C, W & Infant 1618; A & W 1650.
Ormskirk. A her lg 1500.
Preston. C 1623.
Rivington. Skl lg 1627.
Rochdale. Skl 1668.
Rufford. A pal 1543.
Sefton. L can 1528; A & 2W 1570*; A & 2W 1568.
Ulveston. C & W 1606.
Walton-on-the-Hill. C 1586.
Whalley. A & W kng 1515.
Winwick. A her tr can mt lg wn 1492; A in chasuble & W her wn 1527†.

LEICESTERSHIRE
Aylestone. E lg 1594.
Barwell. E in pulpit, W kng Infant 1614*; C & W 1659.
Bottesford. E Sts tr can B. V. Mary lg 1404††; E mt 1440.
Castle Donington. A & W can 1458†.
Hinkley. L 1490.
Hoby. A mt 1480.
Leicester, Wigston's Hospital. Shr 1540.
Loughborough. C mt & W mt wn 1445; C & W 1480.
Lutterworth. C & W 1418; C 1470.
Melton Mowbray. Heart wn 1543.
Packington. E mt 1530.
Queeniborough. L 1634.
Saxelby. L 1523.
Scalford. C kng 1520.
Sheepshed. A & W 1592.
Sibstone. E, Christ on Rainbow 1532*.
Stapleford. A & W 1492.
Stokerston. A mt & W lg 1467; A & W 1493.
Swithland. L wn 1455.
Thurcaston. E mt can mt 1425.
Wanlip. A & W 1393.
Wymondham. L 1521.

LINCOLNSHIRE
Algarkirk. C & 2W B. V. Mary demi 1498.
Althorpe. E demi 1360.
Ashby Puerorum. A & W 1560; A 1560.
Barrowby. C & W 1479; A mt & W her 1508.
Barton-on-Humber, St Mary. L demi wn 1380; C on winecasks lg 1433†.
Bigby. L 1520; E & W 1632.
Boston, St Botolph. C mt tr can Sts mt lg 1398; E Sts lg wn 1400; C mt & 2W on bracket tr can wn 1400; C & W wn 1470; Can mt wn 1500; C demi 1659.
Broughton. A & W lg 1390†.
Burton Coggles. C 1590; A & W 1620.
Burton Pedwardine. L 1631.
Buslingthorpe. A demi 1310†.
Coates, Great. L 1420; C & W kng Resurrection wn 1503.
Coates-by-Stow. A, W & Infant 1590; A & W kng 1602.
Conisholme. A & W 1515.
Corringham. E & W 1628.
Covenham, St Bartholemew. A 1415.

Croft. A demi 1300†.
Driby. C mt & W kng 1583.
Edenham. St Thomas of Canterbury sm 1500*.
Evedon. C & W kng 1630.
Fiskerton. E 1490.
Gedney. L lg 1390†.
Glentham. L demi wn 1452.
Grainthorpe. Cross lg 1380.
Gunby. A & W can mt 1400†; Justice can mt 1419†.
Hainton. C & W 1435; A, W & Daughter her 1553.
Halton Holgate. L 1658.
Harpswell. A & W 1480.
Harrington. L 1480; A & W kng 1585.
Holbeach. A mt wn 1410; L 1488.
Horncastle. A kng pal 1519; Shr mt wn 1519.
Ingoldmells. C with stilt 1520.
Irnham. A can lg 1390†; A mt 1440.
Kelsey, South. A & W 1410.
Laughton, nr Gainsborough. A tr can lg 1400††.
Lincoln, St Mary-le-Wigford. Cross sm pal 1469.
Lincoln, St Peter-at-Arches. C & W kng 1620.
Linwood. C on woolsack & W can lg 1419††; C on woolsack can lg 1421†.
Mablethorpe. L 1522.
Northorpe. C & 2W 1595.
Norton Disney. A & W with Son A & 2W demi pal 1578.
Ormsby, South. L 1410; A & W can 1482.
Pinchbeck. L kng 1608.
Rand. A mt 1500; L 1590.
Rauceby. E sm 1536.
Scotter. C & W kng 1599.
Scrivelsby. A 1545.
Sleaford. C & W 1521.
Somersby. C kng 1612.
Spalding, Johnson Hospital. C & W kng 1597.
Spilsby. L lg 1391†; A & W tr can wn 1400†.
Stallingborough. A & W her 1509; L 1610.
Stamford, All Saints. C on woolsack & W 1460; C on woolsack & W can mt lg 1465†; L sm 1471; C & W 1475; C & W 1500; E mt wn 1508.
Stamford, St John. C & W 1489; E mt wn 1497.
Stamford, St Mary. C 1684.
Stoke Rocheford. A 1470; A & W 1503.
Tattershall. C wn 1411; E 1456; A mt tr can mt Sts lg 1470; L can Sts lg 1470†; L tr can Sts lg 1470†; E Sts lg 1510†; E 1519.
Theddlethorpe, All Saints. A 1424.
Waltham. L, C & L demi 1420.
Winston. 2L wn 1504.
Winthorpe. C & W 1505; C 1515.
Witham, North. C mt 1424.
Wrangle. C & W 1503.

Middlesex

Acton. C kng 1558.
Ashford. C & W 1522.
Bedfont. C & L kng 1631.
Brentford, New. C & W 1528.
Chelsea. L her 1555; A & W kng 1625.
Clerkenwell, St Mary. Bishop mt 1556.
Cowley. C & W 1502.
Drayton, West. C sm 1520; L 1529; C & W 1581.
Ealing. C & W kng 1490.
Edmonton. 2C & L 1500; C & W 1523.
Enfield. L her tr can 1470†; C & W 1592.
Finchley. L mt wn 1480; L sm 1487; C & W 1609; L 1609; C & 3W 1610.
Fulham. L demi Angels fgn 1529*.
Greenford, Great. E demi 1450; L 1480; E 1521.
Greenford, Little. C & 2W sm 1500; C sm 1590.
Hackney. E 1521; A 1545.
Hadley, Monken. L sm 1442; C & W 1500; L 1504; C & W 1518; C & W 1614.
Harefield. L sm 1444; A & W pal 1537; Serjeant at Law & W 1528; A & W pal 1537; C & W kng 1545.
Harlington. E demi 1419; A & W pal 1545.

Harrow. A sm 1370; A lg wn 1390; E mt 1442; E in ac demi 1460; E Sts mt can mt 1468; C & 3W 1480; A & W 1579; C & W 1592; C & W lg 1600; C & W 1603.
Hayes. E demi 1370; A 1456; A & W mt 1576.
Hendon. C 1515; C & W 1615.
Heston. L in bed Infant Angel & Christ 1581*.
Hillingdon. A & W can 1509; A 1528; C & W 1579; C 1599.
Hornsey, Old Church. Infant 1520.
Ickenham. C 1580; C & W 1582; A & W 1584.
Isleworth. A 1450; Nun 1561; C 1590.
Islington, St Mary. A & W can 1450; A & W pal 1546.
Kilburn, St Mary. Head of Nun, 15th cent.
Kinsbury. C & 2W 1520.
London Churches:
All Hallows Barking. C on woolsack & W Heart 1437; C & W kng 1477; C 1498; Resurrection 1500*; C & 2W 1518; C & W can B. V. Mary fgn 1533†; A & W pal 1546; A & W 1560; C 1591.
Holy Trinity Minories. Child sm 1596.
St Andrew Undershaft. C & W 1539; C & 2W kng 1593.
St Bartholomew-the-Less. C & W wn 1439.
St Dunstan-in-the-West. C & W 1530.
St Helen Bishopgate. C mt & W 1465; E in ac 1482; C & W 1495; E in ac 1500; A 1510; A 1514; L her 1535.
St Katherine Regent's Park. C & W kng 1599.
St Martin Ludgate Hill. C 1586.
St Olave Hart St. 2L kng 1516; C & W 1584.
Westminster Abbey. Cross mt 1270; Bishop mt B. V. Mary tr can mt lg wn 1395; Archbishop can lg wn 1397††; L tr can lg 1399††; A lg 1438; A mt 1483; Abbot tr can 1498†; A 1505; E in ac 1561.
Westminster, St Margaret. C & W kng 1597.
Mimms, South. L wn 1448.
Northolt. A sm 1452; A & W pal 1560; E kng 1610.
Norwood. C 1618; C 1624.
Pinner. Infant pal 1580.
Ruislip. C & W 1574; C & W 1593; C 1600.
Stanwell. E demi 1408.
Teddington. C & W 1506.
Tottenham. C & W 1616; L kng 1640.
Willesden. C & W 1492; C 1505; E 1517; A & 2W 1585; L 1609.

MONMOUTHSHIRE
Abergavenny. C & 3W 1579; E 1631; L & Infant 1637.
Llangattock-nigh-Usk. L kng 1625.
Llanover. 2A 1610.
Matherne. C & W kng 1590.

NORFOLK
Acle. C mt sm 1533; E in ac sm 1627.
Aldborough. A 1481; L 1485; C 1490.
Antingham. A & W 1562.
Attlebridge. Chalice & Wafer 1525.
Aylsham. E wn 1490; C & W sm 1490; 2Shr 1499; C & W wn 1500; Shr 1507.
Baconsthorpe. L her kng 1561.
Barnham Broom. C demi pal 1467; C & W 1514.
Barningham Norwood. A & W 1516.
Barningham Winter. A 1410.
Bawburgh. C 1500; Shr 1505; Chalice & Wafer 1531; Shr 1660.
Beechamwell. E 1385; E 1430.
Beeston Regis. C with whistle & W 1527.
Belaugh. Chalice & Wafer 1508.
Binham. C & W demi 1530.
Bintry. Chalice & Wafer wn 1510.
Blickling. C demi 1360; A lg 1404†; C & W 1454; L 1458; Child 1479; L 1485; L, Infants in arms 1512*.
Brampton. 2Shr B. V. Mary sm 1468; A & 2W 1535; C & W 1622.
Brisley. E mt 1531.
Buckenham, Old. Stork 1500; Chalice & Wafer 1520.
Burgh St. Margaret. E kng 1608.

Burlingham, South. Chalice & Wafer 1540.
Burnham Thorpe. A can lg 1420†.
Burnham Westgate. L 1523.
Buxton. Chalice & Wafer 1508.
Bylaugh. A & W 1471.
Cley. C 1450; C 1460; 2Shr 1512; E sm 1520.
Clippesby. C & W 1503; A & W 1594.
Colney. Chalice & Wafer 1502.
Creake, North. C holding Church tr can lg 1500†.
Creake, South. E demi 1400; E mt & C mt 1470.
Cressingham, Great. A & W 1497; C 1509; E 1518; L mt 1588.
Cromer. L sm 1518.
Dereham. E demi am 1479; L 1486; Goose 1503.
Ditchingham. C & W 1490; 2C 1505.
Dunston. C & 2Shr 1649.
Ellingham, Great. L 1500.
Elsing. A mt can Sts Weepers including Edward III lg 1347††.
Erpingham. A lg 1415†.
Fakenham. E mt wn 1428; C mt & SW 1470; 4 Hearts 1470.
Felbrigg. C & W Son A & W 1380†. A with Garter & W can lg 1416††; L sm 1480; A 1608; L 1608.
Feltwell St Mary. A 1479; L 1520.
Fincham. Shr sm 1520.
Fransham, Great. A can 1414; Shr 1500.
Frenze. A 1475; A 1510; L 1519; Shr sm 1520; L 1521; L mt 1551.
Frettenham. L wn 1420; L 1460.
Gorleston. A mt 1320†.
Guestwick. Chalice & Wafer 1504; C 1505.
Halvergate. L demi pal 1520*.
Harling, West. E 1479; A & W 1490; A & W 1508.
Heacham. A 1485.
Hedenham. Chalice & Wafer 1502.
Heigham. C sm 1630.
Helhoughton. Heart, Hands etc. 1450.
Hellesdon. C & W demi 1370; E 1389.
Hindolvestone. C & W kng 1568.
Holme-next-the-Sea. C & W 1405.
Honing. A 1496.
Hunstanton. C & W 1480; A her tr can Ancestors her bracket lg 1506††.
Ingoldisthorpe. E, W mt & Daughter 1608.
Ketteringham. L 1470; A & W her kng 1499; Infant 1530.
Kimberley. A & W 1530.
Kirby Bedon. Heart 1450; 2Shr sm 1505.
Langley. C 1628.
Loddon. Heart, Hands etc. 1462; 2Shr 1546; A her 1561; C & W 1609.
Lynn, St Margaret. C & W can Sts Souls etc. fgn wn (plate 118″ x 68″) 1349††; C & 2W can Weepers Souls etc. fgn wn (plate 107″ x 61″) 1364††.
Lynn, West. E 1503.
Martham. Heart 1487.
Mattishall. C 1480; C 1510; C & W 1510.
Merton. A & 2W her kng 1495; A mt pal 1562.
Methwold. A her mt can mt 1367.
Metton. C & W demi 1493.
Mileham. C & W 1526.
Morston. E 1596.
Morton-on-the-Hill. L 1611.
Narborough. C & W 1496; A 1545; Justice & W her kng Resurrection 1545; A & W kng 1561; A 1581.
Necton. L 1372; L lg 1383; Notary & W 1499; C sm 1528; C & W 1532; L 1596.
Newton Flotman. 3A kng 1571.
Norwich Churches:
St Andrew. Can mt 1467; C & W 1500.
St Clement. L 1514.
St Etheldred. E mt 1485.
St George Colegate. C & W on bracket 1472.
St George Tombland. C 1450.
St Giles. C & W 1432; C & W 1436; Chalice & Wafer 1499.
St John Maddermarket. C & W wn 1412; C sm 1450; C & W 1472; C & W 1476; L 1506; C & W on brackets 1524; C & W on bracket 1525; C & W on bracket pal 1558.
St John de Sepulchre. C & W wn pal reverse Monk behind bars can 1535*; A & L 1597.

St Laurence. C mt 1436; C 1436; Prior on bracket can mt 1437†; Skl 1452; E 1483; C 1495.
St Margaret. L 1577.
St Michael-at-Plea. Skl 1588.
St Michael Coslany. 2Shr 1515.
St Peter Mancroft. A pal 1568.
St Stephen. L 1410; C wn 1460; C & W 1513; 2C 1513; E 1545.
St Swithin. C & W 1495.
Ormesby, Great. L demi 1440; A 1529.
Outwell. A 1511.
Paston. C pal 1570.
Plumstead, Little. C mt W 1480; A 1565.
Rainham, East. C 1500; E 1522.
Raveningham. L 1483.
Reedham. L 1474.
Reepham. A mt & W can mt lg 1391†.
Ringstead, Great. E 1482.
Rougham. A in legal robes & W 1472; A & W 1510; 2 Infants can sm 1510; C & 2W 1586.
Sall. C mt 1420; C & W 1440; C & W on bracket mt 1441; 2L 1453; Shr 1454.
Salthouse. Chalice & Wafer 1519.
Scottow. Chalice & Wafer wn 1520.
Sculthorpe. A kng 1470; C & W 1521.
Sherrington. A 1445; E wn 1486; L 1520; A mt & W kng 1593.
Shernbourne. A & W 1458†.
Shotesham St Mary. A & W 1528.
Snettisham. L 1560; C & W 1610.
Snoring, Great. Head of A & W 1424.
Southacre. A & W her lg 1384†; Heart, Hands etc. 1454; E kng B. V. 1534.
Sparham. E 1490.
Sprowston. Z mt & W kng 1559.
Stalham. C & W sm 1460.
Stokesby. A & W 1488; E in ac mt 1506; L 1570; L 1614.
Stradsett. A 1418.
Surlingham. E in ac 1460; Chalice & Wafer mt 1513.
Swaffham. A wn 1480.
Swanton Abbot. E 1477.
Thwaite. C & W 1469.
Tottington. L 1598.
Trowse. L 1585.
Trunch. Heart 1530.
Tuddenham, East. C & 2W 1500.
Tuddenham, North. Cross sm 1625.
Upwell. E tr can 1430†; E 1435.
Walsham, North. Chalice & Wafer 1520.
Walsingham, New. C & W 1485; C & W 1509; Chalice & Wafer 1520.
Warham All Saints. A sm 1474.
Weston. L 1533.
Whissonsett. A 1484; A 1484.
Wiggenhall St Mary-the-Virgin. Heart 1450.
Witton. L sm 1500.
Wiveton. E 1512; Shr 1540; C & W 1597.
Wood Dalling. E 1465; C sm 1504; C sm 1507; Chalice & Wafer 1510; 2C demi 1518.
Woodton. L sm 1532.
Worstead. E demi 1404; C 1500; E demi sm 1520.
Yelverton. L demi 1525.

NORTHAMPTONSHIRE

Addington, Great. E 1519.
Aldwinkle All Saints. C 1463.
Ashby, Canons. C 1584.
Ashby, Castle. E Sts lg 1401†.
Ashby St Ledgers. C & W can sm 1416; Shr mt 1471; A & W her can 1494†; A her kng 1500; A her 1553; E 1510.
Ashton. C & W 1584.
Aston-le-Walls. C & 2W kng 1609.
Barnwell St Andrew. C & W kng sm 1610.
Barton, Earls. C & W 1512.
Barton Segrave. L kng Infant 1615.
Blakesley. A 1416.
Blatherwyck. A & W mt 1548.
Blisworth. A & W 1503.
Boddington. E 1627.
Brampton, Church. Skl wn 1585.
Brompton-by-Dingley. A & W mt can mt 1420; A 1476.
Brington, Great E demi on bracket 1340.
Burton Latimer. L & Infant 1626.
Chacombe. Trinity 1545.
Charwelton. A & W 1490; C & W can lg 1490; A & W pal 1541.

Chipping Warden. E sm 1468; C & W mt 1584.
Collyweston. L sm 1508.
Cotterstock. E on bracket can 1420†.
Cranford St Andrew. A & W 1418; C & 2W 1602.
Cransley. C & W 1515; A & W kng 1589.
Dene. A & 2W kng 1584; A & W 1586; A & W 1606.
Dingley. L kng 1577.
Dodford. A & W 1414; A & W 1422; L 1637.
Easton Neston. A & W 1552.
Farndon, East. E 1622.
Fawsley. A her Heart 1516; A & W 1557.
Floore. A & W 1498; A & W 1510; Cross sm mt 1537.
Geddington. C & W mt 1480.
Grendon. L & 2A 1480.
Harrington. A & W kng 1545.
Harrowden, Great. A & W on brackets 1433.
Hemington. C & W 1517.
Heyford, Nether. A & W lg 1487.
Higham Ferrers. E can Sts mt lg 1337††; Cross 1400; C & W can lg 1425; L mt 1435; E 1498; C & W 1504; Heart 1500; C sm 1518; E 1523; C 1540; C 1540.
Holme Pierrepont. L lg 1390.
Horton. A & 2W 1491.
Irchester. L 1510.
Kettering. A & W kng 1630.
Lowick. A & W 1467.
Marholm. A & W her kng 1534.
Naseby. C mt & W 1446.
Newbottle. C & 2W kng 1555.
Newnham. L 1467.
Newton Bromshold. E sm 1426; E 1487.
Newton-by-Gedding. C & W kng Cross & St Faith 1400†; L 1604.
Northampton, St Sepulchre. C & 2W lg 1640.
Norton. C & W 1504.
Norton, Green's. A & W lg 1462; L 1490.
Orlingbury. C & W 1502.
Potterspury. L sm 1616.
Preston Deanery. A & W 1622.
Raunds. C & W 1510; L 1510.
Rothwell. E Angels 1361; C & W 1540; C kng 1591.
Spratton. C mt & W wn 1474.
Staverton. C & W kng 1590.
Stoke Bruerne. E kng 1625.
Sudborough. C & W 1415.
Sulgrave. C mt 1564.
Tansor. E 1440.
Wappenham. A mt 1460; Justice & W mt 1479; L sm 1499; A & W sm 1500; A & W sm 1500.
Warkworth. A mt 1412; A 1420; L 1420; L 1430; A 1454.
Welford. A & 3W kng 1585.
Woodford-cum-Membris. E 1420.
Woodford, nr Thrapstone. A 1580.

NORTHUMBERLAND
Newcastle-upon-Tyne, All Saints. C & W tr can Sts Souls etc. fgn lg 1411††

NOTTINGHAMSHIRE
Annesley. C with Bow, Arrows & Hound 1595*.
Clifton. A 1478; A 1491; C & W 1587.
Darlton. A & W 1510.
Hickling. E 1521.
Markham, East. L 1419.
Newark. C tr can Weepers Soul etc. (plate 112" x 68") 1361††; C 1540; C 1557.
Ossington. A & W pal 1551.
Radcliffe-on-Trent. L 1626.
Stanford-on-Soar. E mt 1400.
Strelley. A & W 1487.
Wollaton. A & W 1471.

OXFORDSHIRE
Adderbury. A & W 1460; L 1508.
Alvescot. C & W 1579.
Aston Rowant. C & W sm 1445; C & W 1470; L mt sm 1508.
Bampton. E demi 1420; E sm 1500; L 1633.
Barford, Great. C & W mt 1495.

Beckley. L kng 1619.
Bicester. C & W 1510.
Brightwell Baldwin. Justice & W kng sm 1439; Justice & W can lg 1439††.
Brightwell Salome. E sm 1492.
Broughton. L lg 1414†.
Burford. C & W kng bracket mt 1437; C kng 1609; C & W kng 1614.
Cassington. Cross 1414; Shr 1590.
Caversfield. C 1435; L 1506; Heart Hands etc. 1533.
Chalgrove. A 1441; A & 2W 1446.
Charlton-upon-Otmoor. E 1476.
Chastleton. L sm 1592; C & W mt sm 1613.
Checkendon. Serjeant-at Law tr can mt 1404†; Soul & Angels 1430*; L 1490.
Chesterton. C & W 1612.
Chinnor. Head of E in Cross 1320†; E in ac demi lg 1361; A & 2W lg 1385; A & W demi 1385; E demi lg 1388; L demi 1390; A lg 1392; C mt & W mt 1410; C 1410; A mt 1430; C & W 1514.
Chipping Norton. C & W sm wn 1450; C on woolsack & W lg 1451; C mt 1460; C mt & W 1484; L 1503; L 1507; L 1530.
Cottisford. A & W kng 1500.
Crowell. E demi 1469.
Cuxham. C & 2W sm 1506.
Deddington. C demi 1370.
Dorchester. A mt lg 1417; L sm 1490; Abbot 1510†; C & W mt 1513.
Ewelme. A & W 1436†; E demi 1458; E demi 1467; E demi 1498; E in ac 1517; Serjeant-at-Arms & W 1518; C & W kng 1599.
Fifield. L kng 1620.
Garsington. C & W sm wn 1484.
Glympton. C 1610.
Goring. L can 1401; C & W pal 1600.
Hampton Poyle. A & W 1424.
Handborough. L & 2C wn 1500; Shr 1567.
Harpsden. L sm 1460; A & W 1480; E sm 1511; L 1620.
Haseley, Great. E 1494; Shr 1497; L 1581.
Heythrop. A & W 1521.
Holton. A 1461; Child 1599.
Hornton. C 1586.
Ipsden. A & W pal 1525.
Islip. L & 2C 1637.
Kiddington. E in ac sm 1513.
Kingham. L kng 1588.
Lewknor. E demi 1380.
Mapledurham. A can lg 1395†.
Middleton Stoney. L 1607.
Milton, Great. L 1510.
Newnham Murren. L kng mt 1593.
Noke. L, C & Justice kng 1598.
Northleigh. A lg wn 1431.
Nuffield. C demi 1360.
Oddington. Skl eaten by Worms 1510*.
Oxford Colleges:
All Souls'. E kng mt sm 1461; E in ac 1490; E in ac & C demi 1510.
Christ Church. C sm 1450; C mt sm 1452; E 1557; E in ac kng 1578; E in ac kng 1584; E in ac demi 1587; C kng 1588; C kng 1602; E in ac kng 1613.
Corpus Christi. Skl 1530; C 1602.
Magdalen. E in ac demi 1478; E in ac 1478; E 1480; E in ac 1480; E in ac 1480; E in ac demi 1480; E 1487; E in ac demi 1500; E in ac 1500; E in ac 1501; E in ac demi mt 1502; E 1515; C in ac 1523; E with Garter pal 1558.
Merton. E demi lg 1311; E in ac sm in Cross mt 1370; 2E in ac on bracket can 1420†; E in ac demi 1445; E Sts tr can mt lg 1471†; E in ac demi 1519.
New College. E 1403; Archbishop tr can lg 1417††; E demi 1419; E in ac 1427; E in ac 1441; E in ac 1447; E in ac demi 1451; E in ac holding Cross 1468; Shr sm 1472; E in ac holding Cross mt 1478; E in ac 1479; E 1494; E demi 1507; E in ac 1508; Notary 1510; E 1521; Bishop mt lg 1525; E in ac 1592; E in ac kng 1601; E in ac 1619.
Queen's. E in ac wn 1477; E 1518; Bishop kng 1616*; E in ac kng 1616*.

St John's. E in ac kng 1571; E in ac kng 1573; E in ac kng 1577; E in ac kng 1578.

Oxford Churches:

St Aldgate. 2 C kng 1607; C kng 1612; E in ac kng 1637.

St Cross Holywell. L in bed Infant 1622*; 2L kng 1625.

St Mary Magdalen. E in ac kng 1580.

St Mary-the-Virgin. E kng St Katherine 1507; C kng 1581; C kng 1584.

St Michael. C & W kng 1578; C in pew 1617.

St Peter-in-the-East. C & W sm wn 1478; C & W kng 1572; C & 2W pal 1574; C & W kng 1599.

St Peter-le-Bailey. C mt 1419; L 1420; C & W kng 1650.

Pyrton. C & W sm 1522.

Rollright, Great. E 1522.

Rotherfield Greys. A can lg 1387†.

Shiplake. C & W 1540.

Shipton-under-Wychwood. Shr pal 1548.

Shirburn. A & W kng 1496.

Somerton. A & W 1552.

Souldern. Heart Hands etc. wn 1460; E sm 1514; C 1580.

Southleigh. C 1557.

Stadhampton. C & W 1498; C & W 1508.

Stanton Harcourt. 2C sm 1460; L sm 1516; L 1518.

Steeple Aston. C & W 1522.

Stoke Lyne. C & W kng Christ 1535; C & W kng 1582.

Stoke, North. E with Garter demi mt 1363.

Stoke Talmage. C & W sm 1504; A & W 1589.

Swinbrook. A & 3W 1470; A her 1510.

Tew, Great. A & W can lg wn 1410†; Trinity 1487; C & W 1513.

Thame. A, W, Son A mt & W on brackets mt 1420; A & W mt 1460; C & W wn 1500; C & 2W 1502; C & W 1503; C & W wn 1508; A her kng 1539; C mt 1543; C kng 1597.

Waterperry. L sm 1370; A & W pal 1540; A 1530.

Watlington. C & W 1485; 2Shr 1501; C 1588.

Whitchurch. A & W 1420; E 1455; E kng 1610.

Witney. C & 2W Trinity 1500; C 1606.

Woodstock. C sm 1441; E in ac kng 1631.

RUTLAND

Braunston. C & W 1596.

Casterton, Little. A & W 1410†.

Lyddington. L 1486; C & W 1530.

SHROPSHIRE

Acton Burnell. A can lg 1382†.

Acton Scott. C & W kng 1571.

Adderley. Abbot mt lg 1390; A & W 1560.

Alveley. C 1616.

Burford. L lg 1370.

Drayton. C kng 1580.

Edgemond. Shr & W 1533.

Glazeley. C & W 1599.

Harley. A & W 1475.

Ightfield. L tr can St. John Bapt. lg 1495†; C lg 1497.

Myddle. C & W 1564.

Plowden Hall, Private Chapel. C & W 1557.

Tong. A & W 1467††; E wn 1510; E in ac 1517.

Upton Cresset. C & W kng 1640.

Wenlock, Much. C & W kng 1592.

Withington. A mt & W 1512; E 1530.

SOMERSETSHIRE

Axbridge. C & W kng 1493.

Blackwell. C & W kng 1604.

Bagborough, West. Heart 1641.

Banwell. C & W sm 1480; E 1503; C mt 1554.

Batcombe. E kng 1613.

Bath Abbey. A & W kng 1639.

Beckington. A & W 1485; C & W Angels 1505.

Burnett. C & W kng 1575.

Cheddar. A 1442; L 1475.

Chedzoy. A 1490.

Churchill. A & W 1572.

Combe Florey. L 1485; A 1526.
Çombe Hay. A 1669.
Cossington. A & W 1524.
Crewkerne. A kng sm 1525.
Croscombe. C & W kng 1606; C & W kng 1625.
Dowlish Wake. A 1528.
Dunster. C & W 1520.
Fivehead. L pal 1565.
Hinton St George. C & W 1590.
Hutton. A & W 1496; A & W kng 1528.
Ilminster. A & L tr can lg 1440†; A & W lg 1618††.
Ilton. Infant 1508.
Kittisford. A & W 1524.
Langridge. L 1451.
Luccombe. C 1615.
Lydiard, Bishop's. C & W kng 1594.
Minehead. L mt can mt 1440.
Petherton, North. L kng 1652.
Petherton, South. A & W can lg 1430†; L wn 1442.
Portbury. L kng Infants 1621.
St Decumans. A & W 1571; A & W lg 1596†; A 1616.
Shepton Mallet. A & W kng Death etc. 1649.
Stogumber. L 1585.
Swainswick. C 1439.
Thorne St Margaret. C 1600.
Tintinhull. E demi 1464.
Weare. C 1500.
Wedmore. Heart 1630; C with pike & sword 1630.
Wells Cathedral. E demi 1460; C kng 1618*.
Wells, St Cuthbert. C kng sm 1623.
Yeovil. Friar demi 1460; C & W 1519.

STAFFORDSHIRE
Audley. A lg 1385†; C 1628.
Biddulph. C & W 1603.
Blore-Ray. C mt & W 1498.
Bromley, Abbot's. C demi 1463.
Clifton-Campville. L demi on bracket mt pal 1360.
Eccleshall. L kng sm 1672.
Hanbury. E wn 1480.
Horton. C & W kng 1589.
Kinver. A & 2W 1528.
Leek. C & 3W kng 1597.
Madeley. C & W 1518; Schoolboy kng sm 1586.
Norbury. L 1360†.
Okeover. A & 2W can 1447†.
Rugeley, Old Church. C 1566.
Standon. Cross sm 1420.
Stone. A & W 1619.
Trentham. A & W kng 1591.

SUFFOLK
Acton. A lg 1302?††; L tr can mt lg 1435†; A 1528; C & W sm 1589; C sm 1598.
Aldeburgh. C mt 1519; L 1520; L 1570; C & W 1601; C & W 1606; C 1612; C & 2W 1635.
Ampton. C & W kng 1480; L 1480; L pal 1490.
Ash Bocking. A & 2W 1585.
Assington. A & W 1500.
Barham. C & W 1514.
Barningham. E in ac sm 1499.
Barrow. A & 2W kng 1570.
Barsham. A lg 1415.
Belstead. A & 2W 1518.
Benhall. C & W 1598; A & W 1611.
Bergholt, East. C 1639.
Bildeston. L 1599.
Boxford. Infant in cot 1606*.
Bradley, Little. C & W kng 1510; A mt 1530; C & 2W kng Infants 1584; C & W 1605; A & W kng 1612.
Braiseworth, Old Church. A 1569.
Bredfield. C & W 1611.
Bruisyard. 2L 1611.
Brundish. E 1360; A 1559; A & W 1560; L mt 1570.
Burgate. A & W can lg 1409†.
Bury St Edmund's, St Mary. C & W kng 1480; E 1514; Tau Cross sm 1520.
Campsey Ash. E can mt 1504.
Carlton. C 1480; C 1490.
Chattisham. L 1592.
Cookley. C & W 1595.
Cowlinge. C & W 1599.

Debenham. A & W demi 1425.
Denham, nr. Eye. C pal 1574.
Denstone. A & W her 1524; L 1530.
Depden. L (twice) & 2A kng 1572.
Easton. A 1425; A 1584; L 1601.
Edwardstone. C & W 1636.
Ellough. L sm 1520; L sm 1607*.
Elmham, South, St James. C & W 1500.
Euston. C & W 1480; C mt & W mt 1520; L sm 1520; A mt & W 1530.
Eyke. Baron of Exchequer mt & W 1430; E 1619.
Fornham All Saints. C mt 1599.
Fressingfield. A & W 1589.
Gazeley. Chalice 1530.
Hadleigh. L 1593; C kng 1637; C & W demi 1637.
Halesworth. C demi 1576.
Hawkedon. C & W wn 1510.
Hawstead. Boy 1500; Girl 1500; L 1530; A & 2W 1557.
Holbrook. A 1480.
Honington. C 1594.
Ipswich Churches:
St Clement. C & 2W wn 1583; C 1607.
St Mary Quay. C & W fgn 1525†; C & W kng 1565; L 1583. (First Brass now in Christ Church Museum.)
St Mary Tower. Notary lg 1475†; C & 2W 1500; L, Notary & C 1506; C & 2W on bracket mt 1520.
St Nicholas. C & W 1475; C 1490; C & W 1600.
St Peter. C & W 1604.
Ixworth. C & W kng 1567.
Kenton. A & W her kng Crucifix 1524.
Kettleburgh. C & 2W mt 1593.
Knodishall. A & W 1460.
Lakenheath. C & W sm 1530.
Lavenham. 2Shr 1486; C & W kng 1560; Infant 1631.
Letheringham. A her lg 1389†; A 1480.
Lidgate. E sm 1380.
Lowestoft. 2Skl mt 1500; C & W mt 1540.
Melford, Long. L 1420; C 1420; L her can mt 1480; L 1480; A 1577; C & 3W Infants 1624.
Melton. C mt W & E tr can mt 1430.
Mendham. L 1615; C 1616; C 1634.
Mendlesham. A 1420.
Mickfield. C & W sm 1617.
Middleton. C & W sm 1500; C 1610.
Mildenhall. A 1618.
Monewden. E in ac kng 1595.
Nayland. L mt can mt 1485; C & W can mt 1500; C & W wn 1516.
Nettlestead. A sm 1500.
Occold. C & W 1490.
Orford. C sm 1480; L sm 1490; L & 2C sm 1500; C sm 1510; C sm 1510; L sm 1510; C & W Trinity sm 1520; C sm 1520; C pal 1580; C & W 1591; L 1605; C & W kng 1640.
Pakefield. C & W wn 1417; E in ac demi 1451.
Petistree. C & 2W 1580.
Pettaugh. C & W 1530.
Playford. A her lg 1400†.
Polstead. E 1430; C & W 1490.
Raydon. E mt sm 1479.
Redgrave. L 1609.
Rendham. Chalice 1523.
Ringsfield. A & W her kng 1595.
Rougham. A & W lg 1505†.
Saxham, Great. C 1632.
Sibton. C & W kng 1574; C mt & W 1582; C & W kng 1626.
Sotterley. A & W 1479; A 1480; L kng 1630; A kng 1630; A 1572; L 1578.
Southolt. L 1585.
Spexhall. L 1593.
Stoke-by-Clare. L 1530; C 1597; L 1605.
Stoke-by-Nayland. L 1400; A lg 1408†; L 1535; L & Infant 1632.
Stonham Aspall. E 1606.
Stowmarket. Shr 1638.
Stratford St Mary. C & W 1558.
Tannington. L 1612.
Thurlow, Great. A & W 1460; L mt 1460; A mt & W 1530.
Thurlow, Little. A & W 1520.
Ufford. C mt & 3W 1488; Skl 1598.
Waldingfield, Little. C & W 1506; A & W 1526; L 1530; C 1544.
Walton. C & W 1459; Child sm wn 1612.

Wenham, Little. A & W can 1514.
Wickhambrook. C & 2W 1597.
Wickham-Skeith. L kng 1530.
Wilby. C, Sheep 1530.
Woodbridge. Child 1601.
Worlingham. C & W 1511.
Wrentham. L 1400; A 1593.
Yaxley. C 1598.
Yoxford. A & W 1428; Shr lg 1485*; C 1613; L 1618; L 1618.

SURREY

Addington. A 1540; C & W 1544.
Albury, Old Church. A wn 1440.
Barnes. 2L sm 1508.
Beddington. L 1414; Cross sm 1425; C & W can lg 1432†; C & W 1430; A 1437; L sm 1507; A & W her 1520.
Betchworth. E 1533.
Bletchingley. L sm 1470; E 1510; C & W Trinity 1541.
Bookham, Great. L sm wn 1433; L 1597; C & W 1598; C 1668.
Byfleet. E 1480.
Camberwell, St Giles. C sm wn 1497; C kng mt 1499; A & W kng 1532; A 1465; C & W kng 1570; C & W kng 1577.
Carshalton. A & W her 1485†; E mt 1493; A mt can B. V. Mary 1497; L 1524.
Charlwood. A & W kng 1553.
Cheam. C mt 1390; C demi 1390; C & W demi 1458; C demi 1459; A 1475; C & W lng Trinity pal 1542.
Chipstead. L 1614.
Clapham, St Peter. E 1470.
Cobham. Nativity sm 1500*. A pal 1550.
Compton. C & W wn 1508.
Cranley. Resurrection 1503*; E demi sm 1510.
Crowhurst. A 1450; A 1460.
Croydon, St John. E 1512; A & W mt 1562.
Ditton, Long. C & W 1527; C & W 1616.
Ditton, Thames. A & W kng 1559; L & 2C 1580; C & W kng 1582; C & W kng & C & W kng 1587; C & W 1590.
Egham. C & 2W kng 1576.
Ewell. L her 1519; L wn 1521; L pal 1577.
Farley. C & W sm 1495.
Farnham. C & W kng 1494; L kng 1597.
Godalming. C & W 1509; A 1595.
Godalming, Wyatt's Almshouses. C & W kng 1619.
Guildford, Holy Trinity. C 1500; C & W kng 1607.
Guildford, St Mary. C & W 1500.
Horley. L can lg 1420†; C 1510.
Horsell. C 1603; C 1603; C & W 1619.
Horsley, East. C demi 1390; Bishop kng sm 1478*. C & W 1498.
Kingston-upon-Thames. C & W 1437; C mt & W kng 1488.
Lambeth, St Mary. L her 1535; A 1545.
Leatherhead. C mt wn 1470.
Leigh. C & W 1449; L sm 1450; Trinity 1499.
Lingfield. L lg 1375†; A lg 1403†; A 1417; L can 1420†; L demi 1420; E mt wn 1440; E demi 1445; L sm 1450; E demi sm 1458; E 1469; E 1503.
Merstham. C & W sm 1463; 2L sm 1473; A 1498; A & W mt 1507; Child & Infant sm 1587.
Mickleham. C & W kng 1513.
Mortlake. Child 1608.
Nutfield. C & W 1465.
Oakwood. A sm 1431.
Ockham. E demi 1376; A & W 1483.
Oxted. E mt 1428; L 1480; Child mt 1611; Child 1611; Child 1613.
Peper-Harrow. L kng Trinity 1487; Cross 1487; L sm 1621.
Putney. A 1478; L mt wn 1585.
Puttenham. E 1431.
Richmond. C & W kng 1591.
Rotherhithe. C & 2W wn 1614.
Sanderstead. C & W sm 1525.
Send. C & W sm 1521.
Shere. E 1412; C 1512; C & W 1516; L 1520; A 1525.
Stoke D'Abernon. A lg 1277††; A can mt lg 1327††; L 1464; Infant 1516; C & W kng 1592.
Streatham. E 1513.

Thorpe. C & W 1578; C & W kng 1583.
Titsey. C & W kng 1579.
Tooting Graveney. C & W kng 1597.
Walton-upon-Thames. C in Hunting-dress & W, Device of Man riding Stag pal 1587*.
Wandsworth. Serjeant-at-Arms mt wn 1420.
Weybridge. 3Skl 15th cent.; C & 3W kng 1586; C & 2W 1598.
Witley. C & W 1530.
Woking. L 1523; C & W 1527.
Wonersh. C & W 1467; C & W 1503.

SUSSEX

Amberley. A her 1424.
Angmering. L 1598.
Ardingly. C & W can 1500; A & W can 1504; A & W 1510; L 1633; Child 1634.
Arundel, Fitzalan Chapel. E demi 1382; E 1419; Horse sm 1421*; A mt & W can mt lg 1430; E demi 1450; E 1455; A wn 1465; E demi 1474.
Battle. A wn 1426; E wn 1430; E demi 1435; L 1590; E in ac 1615.
Billingshurst. C & W 1499.
Bodiam. A her mt 1360; Shr sm 1513.
Brede. A mt & W 1493.
Brightling. C & W 1490; Child 1592.
Broadwater. E can 1432†; Cross 1445.
Burton. A kng 1520; L her kng 1558*.
Burwash. C wn 1440.
Buxted. E demi in Cross 1408†; E wn sm 1485.
Chichester Cathedral. Heart, Hands etc. wn 1500; C & W kng 1592.
Clapham. A & W her Trinity 1526; A & W kng 1550; A & W kng 1592.
Clayton. E 1523.
Cowfold. Prior tr can Sts lg 1433††; C 1500.
Crawley. L 1520.
Cuckfield. A 1589; A & W kng 1601.
Etchingham. A mt lg 1388; A, W & Son A tr can lg 1444††; 2L 1480.
Ewhurst. C kng 1520.
Firle, West. A & W 1476; A & W 1595; A & W 1595; A 1595; A & 2W 1595; Shr 1638.
Fletching. A her & W can lg 1380†; Gloves 1440*.
Framfield. C & W kng 1595.
Friston. C & W sm 1542.
Goring. A & W 1490.
Grinstead, East. 2A 1505; C sm 1520.
Grinstead, West. L can mt 1440; A & W can lg 1441†.
Hastings, All Saints. C & W 1520.
Hastings, St Clement. C 1563; C 1601.
Hellingly. L lg 1440.
Henfield. C sm 1559; L & Grandchild 1633*.
Horsham. E mt 1411; L sm 1513.
Horsted Keynes. L mt 1420.
Hurstmonceux. A can lg 1402†.
Iden. E 1427.
Isfield. A & W 1558; A & W 1579.
Lewes, St Michael A mt 1430; E demi 1457.
Northiam. E 1518; C 1538.
Ore, St Helen. C & W can mt 1400.
Poling. E demi 1460.
Pulborough. E can wn 1423; C & W wn 1452; C 1487.
Rusper. C & W demi 1370; C & W 1532.
Rye. C mt & W mt 1490; C 1607.
Shoreham, New. C & W 1450.
Slaugham. A can 1503; A & 3W kng Resurrection 1525; L kng 1586.
Slinfold. C & W 1533; L mt 1600.
Stopham. C & W wn 1460; A & W wn 1460; C & W 1462; C kng 1630; C & W 1601; A & 2W 1630; several Children & Shields added c. 1630.
Storrington. E in ac kng wn 1591.
Thakeham. L 1515; C 1517.
Ticehurst. A lg 1370 & 2L sm 1503.
Trotton. L lg 1310††; A with Garter & W can lg 1421††.
Uckfield. C 1610.
Warbleton. E can mt Pelican in Piety lg 1436†.
Warminghurst. C & W 1554.
Willingdon. A 1558.

Winchelsea. C mt 1440.
Wiston. A lg 1426†.

WARWICKSHIRE
Astley. L mt 1400.
Aston. Justice mt & W 1545.
Baddesley-Clinton Hall. L her kng 1520.
Baginton. A & W her lg 1407†.
Barcheston. E in ac 1530.
Barton. C 1608.
Chadshunt. C kng 1613.
Coleshill. E sm 1500; L 1506; E 1566.
Compton Verney. L 1523; A & W 1526; A 1630.
Coughton. A & W 1535.
Coventry, Holy Trinity. C & 2W 1600.
Eatington, Lower. C & W 1603.
Exhall, nr. Alcester. A & W 1590.
Hampton-in-Arden. C sm 1500.
Haseley. A & W pal 1573.
Hillmorton. L 1410.
Itchington, Long. C & 2W kng 1674.
Merevale. A & W lg 1413†.
Meriden. L 1638.
Middleton. Justice & W 1476.
Preston Bagot. L mt 1635.
Shuckburgh, Upper. L mt 1500; A & W 1549; Head of A & W 1594.
Solihull. C & 2W 1549; C & W kng 1610.
Sutton Coldfield. L 1606; C 1621.
Tanworth. L kng 1614.
Tysoe. E sm 1463; L demi 1598.
Upton. E & W kng 1587.
Warwick, St Mary. A & W her lg 1406†.
Warwick, St Nicholas. E 1424.
Wellesbourne-Hastings. A 1426.
Whatcote. E mt sm 1511.
Whichford. E in ac 1582.
Whitnash. C & W 1500; E 1531.
Withybrook. C 1500.
Wixford. A & W can lg 1411††; Child kng 1597.
Wootton-Wawen. A & W 1505.
Wroxhall. L 1430.

WESTMORLAND
Kendal. A 1577.
Musgrave, Great. E 1500.

WILTSHIRE
Aldbourne. C & W 1492; E sm 1508.
Alton Priors. L 1528; C, Angel etc. 1620*.
Barford St Martin. L kng 1584.
Bedwyn, Great. C 1510.
Berwick Basset. C demi 1427.
Blunsden, Broad. L mt 1608; A & 2W 1612.
Bradford-on-Avon. C & W kng Trinity 1520; L 1601.
Brumham. L 1490; A 1516; A & W 1578.
Broughton Gifford. Herald, Death etc. 1620*.
Charlton. C & W 1524.
Chisleden. C & W 1592.
Cliffe-Pypard. A 1380.
Collingbourne Ducis. Child 1631.
Collingbourne Kingston. L 1495.
Dauntsey. A & W 1514; L kng Trinity 1539.
Deane, West. Child 1641.
Devizes, St John. C & W 1630.
Draycot Cerne. A & W 1393†.
Durnford, Great. C & W kng 1607.
Fovant. E in ac kng Annunciation 1500*.
Ham. C & W 1590.
Lacock. A & W her 1501.
Lavington, West. A pal 1559.
Mere. A lg 1398†; A mt lg 1425.
Minety. A & W kng 1609.
Newton, Long. E 1503.
Ogbourne St George. C & W 1517.
Preshute. C mt & W 1518.
Salisbury Cathedral. Bishop demi in Castle lg 1375††; Bishop 1578.
Salisbury, St Thomas. C & W 1570.
Seend. C & W sm 1498.
Stockton. L kng 1590; C & W kng 1596.
Tisbury. C & W 1520; C & W 1590.
Upton Lovell. E demi 1460.
Warnborough. C & W demi 1418.
Warminster. L kng 1649.
Westbury. C & W mt 1605.
Wilton. C & W kng 1585.
Woodford. C 1596.

WORCESTERSHIRE

Alvechurch. A 1524.
Burlingham. C & W kng 1617.
Blockley. E in can kng 1488; E kng 1510.
Bredon. Mitre 1650.
Broadway. A pal 1572.
Bushley. C & W 1500.
Chaddesley Corbett. C & W wm 1511.
Daylesford. C 1632.
Fladbury. A & W 1445; E demi 1458; A 1488; E sm 1504.
Kidderminster. L & 2A tr can lg wn 1415†.
Longdon. A & W 1523.
Mamble. A & W 1510.
Stockton. C 1508.
Stoke Prior. C kng 1606; C & 2W kng 1609.
Strensham. A 1390; A can 1405; A & W 1502; A & W her kng 1562.
Tredington. E 1427; E kng 1482; L 1561.
Worcester, St Helen. C wn 1622.
Yardley. L, C & A 1598.

YORKSHIRE

Aldborough, nr. Borough Bridge (N. & W. R.). A her lg wn 1360†.
Allerton Mauleverer (W. R.). A her & W sm 1400.
Aughton (E. R.). A & W mt 1466.
Bainton (E. R.). E wn 1429.
Beeford (E. R.). E 1472.
Birstall (W. R.). Shr 1632.
Bishop Burton (E. R.). Chalice 1460; L 1521; C & W 1579.
Bolton-by-Boland (W. R.). A & W her kng 1520.
Bossall (N. R.). A mt 1454.
Bradfield, nr. Sheffield (W.R.). C & W kng 1647.
Brandsburton (E. R.) E demi mt wn 1364; A mt & W lg 1397.
Burgh Wallis (W. R.). A mt 1566.
Catterick (N.R.). 2A 1465; A & W 1492.
Cottingham (E. R.). E can lg 1383†; C & W 1504.
Cowthorpe (W. R.). Justice wn 1494.
Everingham (E. R.). L 1558.
Forcett (N. R.) L 1637.
Hampsthwaite (W. R.). C mt sm 1360.
Harpham (E. R.). A & W can mt lg 1418†; A 1445.
Hauxwell (N. R.). C & W kng 1611.
Helmsley (N. R.). wn 1480.
Hornby (N. R.). A mt & W 1489.
Howden (E. R.). A 1480.
Hull, Holy Trinity (E. R.). C & W demi 1451.
Hull, St Mary (E. R.). C & 2W kng 1525.
Ilkley (W. R.). Child 1687.
Kirby Malzeard (W. R.). C & W kng 1604.
Kirkby Moorside (N. R.). L kng 1600.
Kirkby Wharfe (W.R.). E sm wn 1492.
Kirkheaton (W. R.). A & W Infant 1655.
Kirkleatham (N. R.). Child 1628; C 1631.
Laughton-en-le-Morthen (W. R.). A 1620.
Leake (N. R.). C & W sm wn 1530.
Leeds, St Peter (W. R.). A & W 1459; L wn 1467; Chalice 1469; Children 1709.
Lowthorpe (E. R.). A 1417.
Marr (W. R.). C & W 1589.
Otley (W. R.). C 1593.
Owston (W. R.). C & W 1409.
Rawmarsh (W. R.). C & W kng 1616.
Ripley (W. R.). Chalice wn 1429.
Rotherham (W. R.). C & W kng 1561.
Routh (E. R.). A & W can lg 1420†.
Roxby Chapel (N. R.). A 1523.
Sessay (N. R.) E pal 1550.
Sheriff Hutton (N. R.). Infant 1491; L & Infant 1657.
Skipton-in-Craven (W. R.). Trinity 1570.
Sprotborough (W. R.). A & W 1474.
Tanfield, West (N. R.). E 1490.
Thirsk (N. R.). E demi Angels wn 1419.
Thornton Watless (N. R.). Shr 1669.
Todwick, nr. Sheffield (W. R.). C kng 1609.
Topcliffe (N. R.). C & W can Angels fgn lg 1391†.
Wath (N. R.). Justice & W wn 1420; A wn 1490.
Wellick (E. R.). C & W 1621.
Wensley (N. W.). E Angels fgn lg 1375††.
Wentworth (W. R.). A & W kng 1588.
Wilberfosse (E. R.). A & W wn 1447.
Winestead (E. R.). A mt & W mt pal 1540.

Wycliffe (N. R.). Child kng 1606.
York Minster. Archbishop mt lg 1315††; L demi 1585; C demi 1595.
York Churches:
All Saints North St. C demi 1642.
St Crux, Parish Room. C demi 1597.
St Martin Coney St. C demi 1588.
St Michael Spurriergate. Chalice 1466.

IRELAND

Dublin, Christ Church Cathedral. L & Infant sm 1580.
Dublin, St Patrick's Cathedral. E kng 1528; E kng B. V. Mary 1537; C & W 1579.

SCOTLAND

Aberdeen, St Nicholas. E in ac sitting in study, Books, Skulls etc. fgn lg 1613†.
Glasgow Cathedral. A kng 1605.

WALES

Beaumaris (*Anglesey*). C & W kng Trinity Sts 1530.
Bettws Cedewain (*Montgomery*). E 1531.
Clynnog (*Caernarvon*). Child 1633.
Dolwyddelan (*Caernarvon*). A kng 1525.
Haverfordwest (*Pembroke*). C kng 1645.
Holt (*Denby*). Skl 1666.
Llanbeblig (*Caernarvon*). C in bed, Penner Inkhorn etc. 1500*.
Llandough-next-Cowbridge (*Glamorgan*). L 1427.
Llangyfelach (*Glamorgan*). C & W kng 1631.
Llanrwst (*Denby*). A demi 1626; L demi 1632; L demi 1658; A demi 1660; L demi 1669; L demi 1671.
Llanwenllwyfo (*Anglesey*). C & W kng 1609.
Mold (*Flint*). C kng 1602.
Ruthin (*Denby*). C 1560; C & W kng 1583.
Swansea (*Glamorgan*). A & W Resurrection 1500*.
Whitchurch (*Denby*). C & W kng 1575.
Wrexham (*Denby*). Skl 1673.
Yspytty-Ifan (*Caernarvon*). C & W kng Infant 1598.

The Monumental Brass Society

This society has a large number of full members and also a number of associate members who are under eighteen years of age. To become a member one has to be proposed by another member of the society.

The present Honorary Secretary is: John Coales, 90 High Street, Newport Pagnell, Bucks.

The society sends out annually its *Transactions*, which contains interesting articles, and its Portfolio of rubbings.

INDEX

Page numbers in bold type indicate an illustration of a brass at the place named.